Oxford, 2012

The fauna of oil palm and coconut
Insect and mite pests
and their natural enemies

Dominique Mariau

(Translated from the French by Peter Biggins)

Dominique Mariau

The fauna of oil palm and coconut. Insect and mite pests and their natural enemies
Montpellier, France, Cirad-cp

ISBN 2-87614-478-6

CONTENTS

PREFACE 5

LEPIDOPTERA 7

Introduction 7
Tineoidea 7
Yponomeutoidea 11
Gelechioidea 12
Cossoidea 22
Sesioidea 22
Zygaenoidea 22
Pyraloidea 48
Bombycoidea 52
Hesperioidea 53
Papilionoidea 55
Noctuoidea 58
Conclusion 61
Lepidoptera references 63

HEMIPTERA 84

Introduction 84
Hemiptera-Fulgoromorpha 84
Hemiptera-Cicadomorpha 87
Hemiptera-Sternorhyncha 88
Aleyrodoidea 88
Aphidoidea 89
Coccoidea 89
Hemiptera-Heteroptera 92
Depredatory species 92
Disease vectors 94
Predators 95
Conclusion 95
Hemiptera references 97

COLEOPTERA 110

Introduction 110
Scarabaeidae 110
Scarabaeidae references 120
Chrysomelidae 128
Chrysomelidae references 139
Curculionidae 147
Curculionidae references 156
Other Coleoptera 169

Coleoptera on stored fruits 172
Other Coleoptera references 174

ORTHOPTERA 177

Tettigonioidea 177
Acridoidea 178
Grylloidea 180
Orthoptera references 181

PHASMIDA 184

Phasmida references 186

ISOPTERA 188

Termitidae 188
Rhinotermitidae 189
Kalotermitidae 190
Isoptera references 191

THYSANOPTERA 194

Thysanoptera references 195

HYMENOPTERA 196

Hymenoptera references 197

MITES 198

Tetranychidae 198
Tenuipalpidae 199
Eriophyidae (on nuts) 200
Eriophyidae (on leaves) 203
Oribatidae 204
Pyemotidae 204
Mites references 205

GENERAL CONCLUSION 211

INDEX 213

PREFACE

The book by P. Lepesme "Les insectes des Palmiers" is and will remain a reference work for entomologists studying the extensive fauna associated with oil palm and coconut.

Since its publication in 1947, these palms, and primarily the oil palm, have undergone considerable development, firstly in Africa, the birthplace of *Elaeis guineensis,* but also and especially in Asia and Latin America. A little over 50 years ago, there were only around a hundred thousand hectares of oil palm plantings in Asia, compared to the current 7 million worldwide, almost 80% of which are in Asia. They are mostly planted in large estates, but also in small family plots. Half a century ago, there were virtually no oil palms in Latin America, except in the Bahia region of Brazil, where there was a semi-wild grove. Despite numerous phytosanitary problems on this palm in Latin America, it currently grows over an area of around five hundred thousand hectares. Lastly, fifty years ago in West Africa, oil palms were mostly found in wild groves, with varying degrees of human intervention. Cultivated oil palm plantations now cover in Africa around five hundred thousand hectares throughout the continent, and less use is now made of wild groves.

It is therefore not surprising that the fauna associated with *Elaeis*, primarily in its region of origin, has developed in a similar way to the host plant. Many pests, and the fauna associated with them, have adapted perfectly to this new host, switching from indigenous palms or from other plants. Even in Africa, species not seen on oil palm fifty years ago have recently arrived on to the scene. There has also been a substantial amount of research conducted on oil palm pests over that period.

Coconut, for its part, is mostly planted on a multitude of smallholdings, making it difficult to put a figure on the areas planted, but they are estimated at over 8 million hectares, mostly in Indonesia, the Philippines, the Indian sub-continent and the islands of the Pacific Ocean, but also in Africa and Latin America.

Unlike *Elaeis*, the fauna associated with the coconut palm, in its zone of origin (Southeast Asia and the Pacific) has remained much more stable. The fact that the areas planted to *Cocos nucifera* have mostly increased in its zone of origin, where there has been no spectacular development of *Elaeis*, probably has something to do with that. However, other species have adapted to coconut, but in smaller numbers.

Many species are common to both palms. The increase in the number of pest species has gone hand in hand with an increase in the number of parasitoids and predators, about which much more is known than fifty years ago. Whilst this book lists a large number of new species compared to those in Lepesme's book, we have not felt it necessary to include all the secondary, or occasional pests mentioned by that author, notably among the Hemiptera.

In a little over fifty years, all these pests and their enemies have been covered by a thousand notes, papers, theses and articles, of varying importance, in journals related to general entomology or specifically to oil palm and coconut. This book also includes all the observations that an entomologist who has traversed the palm groves of the Tropics for almost forty years could possibly have undertaken, and which have yet to be published or which were merely recorded in reports or notes to which access is difficult or impossible. It is all this knowledge and experience that we felt it would be worth summing up in this work.

The bibliography is not exhaustive either. We felt it would be pointless repeating the lists already provided elsewhere in excellent books such as "Slug and nettle caterpillars on palms in South-east Asia" (M.J.W. Cock, H.C.J. Godefray and J.D. Holloway eds.), among others, or in bibliographical studies on a given species. The references of these publications can be found in the bibliography and are worth consulting for more ample information.

This book is, of course, primarily intended for entomologists and, more generally, to agronomists responsible for the phytosanitary protection of oil palm and coconut plantations, but also for specialists in a given family of insects.

D. Mariau
Montpellier,
November 2000

LEPIDOPTERA

INTRODUCTION

In terms of the number of species, Lepidoptera are the largest group of pests on oil palm and coconut. Around 240 species have been inventoried, belonging to 27 families (the order Lepidoptera contains 124), and 11 superfamilies (38 in the order).

Most Lepidoptera caterpillars, including 95 species belonging to the family *Limacodidae,* are leaf-eaters, attacking unopened (spears) or open fronds. Others are miners of flowers, fruits, stems, nursery seedlings or roots.

All these species have a large number of natural enemies: parasitoids of the different development instars belonging to the orders Hymenoptera and Diptera, themselves parasitized by other species of Hymenoptera, predators among the Hemiptera, Coleoptera, Hymenoptera *Formicidae*, and diseases, with a large variety of viruses and fungi.

TINEOIDEA

PSYCHIDAE

The species of this family are characterized by the construction of a bag at the beginning of the larva cycle, in which the caterpillars develop. The young caterpillars spin a silken thread from which they hang and are driven by the wind. The first larval instar is therefore a dispersal stage for the species, during which the mortality rate is high, which explains the particularly high fecundity (several thousand eggs per female). Only the males have wings. They fertilize females still in their bags, where they also lay their eggs. Females attract males by secreting a pheromone (Rhainds *et al.* 1997). Unlike the species of the sub-family *Stenomatinae*, the caterpillars move around with their bag, which increases in size with larva development. This family includes very serious oil palm pests, particularly in Malaysia.

Mahasena corbetti Tams (Wood 1968, Sankaran 1970, Sankaran and Syed 1972, Syed 1978, Syed and Saleh 1998, Wahid *et al.* 1988, Tiong 1979, Syed *et al.* 1974, Mariau *et al.* 1991).

Large species. The bag can reach 50 mm. The larva cycle (11 to 12 instars) lasts 75 to 82 days, pupation takes one month and incubation 16 days. Major pest on *Elaeis* in Malaysia; 20 host plants have been identified. There is a substantial parasite complex.

Parasitoids

– Diptera *Tachinidae:* These are the major parasitoids: *Eozenillia equatorialis* Townsend on caterpillars, abundant in Sabah, *E. psychidarum* Bar., *Palexorista solennis* (Walker) up to 13% of chrysalis parasitism, *Exorista* (*Zenillia*) *quadrimaculata* Bar. *(Nealsomyia rufella* (Bezzi).

– Hymenoptera *Braconidae*: *Apanteles metesae* Nixon, *A. anyaleti* Muesebeck, *Aulosaphes* sp., *A. psychidivorus* Muesebeck (in Sabah).

– Hymenoptera *Ichneumonidae*: *Echthromorpha agrestoria* Swed. is frequent on the chrysalis. *Xanthopimpla* sp. is much less frequent.

– Hymenoptera *Chalcididae*: *Brachymeria lugubris* Walker on the 3rd larval instar. *B. ghani* Joseph, Nurendran and Joy, on the chrysalis.

– Hymenoptera *Eulophidae*: *Pediobius* sp. (very rare) and *Sympiesis* sp. (very rare on caterpillars).

Predators

– Coleoptera: *Cleridae: Callimerus bellus* Gorh, *C. arcufer* Chapuis (Wood 1968), *Alleculidae Allecula minuta* Pic. and *Allecula* sp., *Phalacridae: Litotursus* sp., *Endomychidae: Sacila variipes* Arrow.

– Hemiptera *Reduviidae*: *Cosmolestes picticeps* Stal. 11 mm, *Isyndus heros* Fabr. 27 mm (rare) and *Sycanus macracanthus* Stal. 15 mm, *Pentatomidae: Cantheconidea* sp.

Diseases

Aspergillus sp. only seen in reared insects.
RNA virus, not included.

Metisa plana Walker (Wahid and Kamaruddin 1993).

Smaller species. The bag only reaches 15 mm. The cycle lasts 3 months, including 50 days for the 4 to 5 larval instars. Rhainds *et al.* (1998) studied the distribution of individuals depending on population densities.

Parasitoids

– Hymenoptera *Braconidae: Dolicogenidea* (*Apanteles*) *metesae* (Nixon) is the major parasitoid on all larval instars. It has several hyperparasitoids: *Paraphylax varius* (Walker), *Pediobius anomalus* (Gahan), which is also a primary parasitoid on the chrysalis, *P. imbreus* (Walker) and *P. lividiscutum* (Gahan), *Eupelmus catoxanthae* Ferrière, also a primary parasitoid, *Eurytoma* sp. and *Aphanogmus hakonensis* Ashmead (*Ceraphronidae*); these last two

are extremely rare. *Aulosaphes psychidivorus* on the final two larval instars and the chrysalis.
.*Ichneumonidae*, *Goryphus bunoh* Gauld. On the final instar caterpillar and the chrysalis (rare), hyperparasitized by *Pediobius* spp.

.*Chalcididae*, *Brachymeria carinata* Joseph, Narendram and Joy on the chrysalis.
.*Eulophidae*, *Tetrastichus* sp. on old caterpillars and the chrysalis (very rare).
.*Elasmidae, Elasmus* sp. caterpillars and chrysalis (extremely rare).
.*Eupelmidae*, *Eupelmus catoxanthae* Ferrière (very rare).

– Diptera *Tachinidae: Eoziniella psychidarum* Bac.

Crematopsyche pendula Joannis (Krishnan 1977). As in the previous case, this is a small insect. The cycle lasts 80 days (5 larval instars, 50 days; eggs, 13 days; pupation, 17 days). The main parasitoids are the *Braconidae, Apanteles psychidivorus* (12%) and *A. metesae* (2%). Caterpillars are attacked by the bug *Sycanus* sp. (*Reduviidae*), itself parasitized by *Gryon* sp. and *Telenomus* sp. (*Scelionidae).*

Pteroma sp.(Wood 1968, Sankaran and Syed 1972). Secondary species in Malaysia. Parasitized by *Parasierola* sp. (*Bethylidae*) and *Brachymeria* sp. (14%), and by *Pediobius lividiscutum* (rare).

Clania tertia Templ. (Sankaran and Syed 1972). Not a very frequent species, probably due to extensive parasitism.
– *Ichneumonidae*, *Xanthopimpla pedator* Fah. and *Xanthopimpla* sp.
– *Braconidae, Aulasaphes psychidivorus.*
– *Tachinidae, Eozenillia equatoralis.* Up to 40% parasitism.

C. bifenestralis Piepers and Snellen, reported on *Elaeis* in Java (Lepesme 1947).

Cryptothelea cardiophaga Westwood. (Sankaran and Syed 1972). Common species controlled by *A. psychidivorus* and the *Ichneumonidae*, *Sericopimpla flavobalteata* Cam. and *Paraphylax rufipes* Cam.

Amatissa cuprea Moore and other species in Malaysia. ***Manatha albipes*** Moore in Sri Lanka (Mathen *et al.* 1962)

Oiketicus kirbyi Guilding (Genty *et al.* 1978, Manolin Avila 1986). A large, common South American species. The caterpillar is 60 mm long. A highly polyphagous species, notably attacking banana. The cycle is very long, from 235 to 320 days, including 200 to 250 days for the 12 to 15 larval instars. Outbreaks are sporadic, but can be

spectacular. Fertility and fecundity are greater in large females (Rhainds *et al.*).
The species is parasitized by *Apanteles* sp., *Elasmus* sp., *Brachymeria* sp., but also by *Conura elaeisis* Delvare (Delvare 1992) and by *Bracon* sp. Parasitism is particularly active in the dry season.

O. abbotti Grote on *Cocos* in Florida and Jamaica (Lepesme 1947), Lastly, ***Eumeta*** sp. on *Cocos* in India. (Dhileepan 1991).

TINEIDAE

This family includes several species on palms, but none is a notable pest, apart from one American species. Some are even detritivorous. Three Asian species are listed by Rapp (1989).

Erechthias psammauba (Meyrick) may attack leaves previously damaged by *Agonoxena.*

E. flaristriata Walsingham. Some damage has been observed on folded leaflets in Tonga, with development of the fungus *Pestalotiopsis.* These caterpillars are parasitized by *Braconidae Meteorus trichogrammae* Wilkinson.

Opogona purpuriella Swezey, also in Tonga on dry coconut tissues.

Hapsifera rugosella Stu (Wood 1968).

Trachycentra elaetropha Meyrick on *Elaeis* fruits in Java and ***T. calamias*** Meyrick on coconut in Fiji (Lepesme 1947).

Tiquadra circumdata Zeller in South America on *Elaeis* (Genty *et al.* 1978). The adult has a wingspan of 40 mm and is white covered with black spots and patches. The caterpillar reaches 32 mm; it is grey and lives protected in an open-ended silken case covered with plant debris. Damage is caused on unopened leaves. However, this species is particularly attracted to decaying organic matter and very ripe oil palm bunches. Populations can be very large (up to 200 caterpillars per frond) causing fronds to dry out. Populations are highly heterogeneous from one palmtree to another. Several Coleoptera (*Staphylinidae* and *Histeridae*) are predators.

YPONOMEUTOIDEA

GLYPHIPTERIGIDAE

Only two species have been reported on oil palm and coconut, one of which is a major pest.

Sagalassa valida Walker (Genty 1973).

Widespread in Latin America. The caterpillar mines oil palm roots near the root mat, killing the downstream section of the attacked root. Final instar caterpillars attack primary roots. On young palms, the root system can be almost completely destroyed, development is halted and the palm topples. The original host plant is a palm of the genus *Bactris*. Attacks particularly develop on the periphery of plantations and seem to be linked to the presence of that palm. The nematode *Steinernema carpocapsae* (Weiser) *Rhabditidae, Steinernematidae* has been tested as a means of biological control. Positive results have been obtained, but it has yet to be proved that the technique is economically viable (Ortiz *et al.* 1994). Chemical control, particularly with Endosulfan, gives significant results. Young roots can be protected, as a preventive measure, by a physical barrier comprising a layer of empty fruit bunches, or fibres at the base of young palms (Andersson 1995).

Iridostoma ichthyopa Meyrick (Lepesme 1947). Reported in Java and Sri Lanka. The caterpillars browse on coconut flowers.

LYONETIIDAE

This family includes a very large number of species originating in the Asia-Pacific zone. All the species, which are generally associated with coconut, though also with oil palm, are detritivorous and therefore only consume dead plant matter (stems-leaves). Some species are acknowledged to be predators.
We shall only list the species belonging to 5 genera, and virtually all inventoried by Lepesme (1947):

Opogona pileigera Meyrick (Africa), ***O. regressa*** Meyrick, ***O. taochroa*** Meyrick, ***O. leucodelta*** Meyrick, ***O. dimidiatella*** Zeller, ***O. aurisquamosa*** Walsingham, to which can be added ***O. trigonomis*** Meyrick (Wood 1968).

Erechthias lampadacma Meyrick.
Dedarchis carpophthora Meyrick, ***D. euophthalma*** Meyrick, in the Solomon islands, ***D. psammaula*** Meyrick in Fiji and Tahiti and

D. heterogramma Meyrick in Fiji (Simmonds 1938), and also ***D. flavistriata*** Walsingham, ***D. heliotoxa*** Meyrick, ***D. minuscula*** Wals, ***D. ophiocypha*** Meyrick, ***D. pachygramma*** Meyrick, ***D. scorpiura*** Meyrick, ***D. spodomicta*** Meyrick.

Ereunetis ochrocrena Meyrick.

Hectacma indicatrix Meyrick.

HELIODINIDAE

Two species on coconut are also detritivorous and can also be predators on scale insects. They belong to the Asia-Pacific region.

Sthathmopoda nucivora Meyrick (Lever 1937) and ***S. adulatrix*** Meyrick (Corbett 1932).

GELECHIOIDEA

The classification of this superfamily has been partially reorganized in line with pre-imago traits (Minet 1989).

ELACHISTIDAE

AGONOXENINAE

Agonoxena argaula Meyrick (Bradley 1965, Hinckley 1961, 1963, Simmonds 1938, Rapp 1989, Singh 1953).

This is the most important species: Fiji– Tonga– Samoa– Futuna– Hawaii– Vanuatu, etc. The adult is small (15 mm). The five larval instars have a life span of 3 weeks. Incubation takes a week and pupation 9 to 10 days. The caterpillar weaves a protective web on the underside of fronds and feeds off the leaflets, sparing the epidermis, eating away small 2 mm wide strips. The status of *A. argaula* on the different host islands was studied by Rapp (1989).
A substantial parasite complex has been observed, including *Braconidae: Agathis* sp. and *Apanteles agonoxenae* Full. on caterpillars in Tonga, *A. orelia* Nixon, in Samoa, but these two species are considered as synonymous (Walker 1989), an *Apanteles* sp. which can parasitize up to 80% of individuals and *Bracon* sp. a *Chalcididae Brachymeria* sp. on chrysalises. In addition, *Macrocentrus* sp. (*Braconidae*) has been introduced into Fiji from Java, *Elachertus agonoxenae* Kerrich (*Eulophidae*) from Papua New Guinea (Kerrich 1961) and *Brachymeria agonoxenae* Fullaway from Java. Lastly, we

would mention *Brachymeria fijiensis* Ferrière, *B. hammari* Crawford and *B. polynesialis* Cameron, which are also hyperparasitoids of *Braconidae* and the Diptera *Tachinidae Tongamyia cinerella* Mesnil. Hymenoptera, especially the *Braconidae*, have numerous hyperparasitoids which limit their effectiveness: *Tetrastichus* sp., *Aphanogmus* (*Ceraphron*) *fijiensis* (Ferrière), *Trichomalopsis* (*Eupteromalus*) sp. (*Pteromalidae*), *Stictopisthus* sp. and *Gelis tenelus* Say (*Ichneumonidae*), *Eupelmus agonoxenae* and *E. cushmani* Crawford, in addition to the *Brachymeria* already mentioned.

A. pyrogramma Meyrick: Solomon islands, Marianne, Guam, Papua New Guinea. The species is parasitized by *Apanteles stantoni* Ashmead and *Macrocentrus* sp. (*Braconidae*), *Ooencyrtus* sp. (Encyrtidae), *Elachertus agonoxenae* Kerrich (*Eulophidae*), introduced into Fiji (Kerrich 1961) and *Actia painei* Grosskey, a small *Tachinidae* (Crosskey 1962).

A. (Haemolytis) miniana (Meyrick): P.N.G., Java. Hawaii (Hinckley 1963) *A. argaula* is parasitized by *Trathala* (*Cremastus*, *Zaleptopygus*) *flavoorbitalis* (Cameron) (*Ichneumonidae*) and *Brachymeria polynesialis* (Cameron). *B. agonoxenae* was successfully introduced from Samoa in 1948, but the introduction of a *Bracon* sp. failed. *Apanteles agonoxenae* introductions also failed.

STENOMATINAE

Three species attack oil palms in South America.
The behaviour of the first two is identical. Eggs are laid primarily in the upper half of the crown. The caterpillars protect themselves in a fixed case made with their excrement. The case increases in size as the caterpillar grows, feeding around its case beneath a web.

Stenoma cecropia Meyrick (Genty 1978).
This highly polyphagous species has a cycle lasting around 3 months; in the final instar, the caterpillar does not exceed 25 mm (8 to 10 larval instars). The case looks like a cornucopia. Very severe outbreaks occur, leading to very serious defoliation (up to 2,000 caterpillars per frond) in Colombia (Magdalena valley). The eggs are parasitized by Hymenoptera *Trichogrammatidae,* including *Trichogramma pretiosum* Riley which parasitizes 2 to 16% of eggs (Calvache, personal communication). Up to 40% of larval instars 5 to 8 are parasitized by *Rhysipolis* sp. (*Braconidae*), and chrysalises by *Brachymeria* sp. *Pseudobrachymeria* (*Chalcididae*) and *Trypoxylon* sp. (*Ichneumonidae*), but such parasitism is only slight.
The predatory activity of Coleoptera *Carabidae*, *Dermestidae* and Hemiptera *Reduviidae* and *Pentatomidae* is low. However, when the ant *Nylanderia fulva* (Mayr) *(Hormiga loca)* is present, no *S. cecropia* outbreaks occur, but it is very aggressive and hinders work in a palm

planting. The caterpillars can also be affected by a polyhedrosis, but it does not seem to be very virulent (Genty, personal communication). A pheromone was discovered (Zagatti *et al.* 1993), but it only seems to play a role at the start of gradation, i.e. when populations are small.

Loxotoma elegans Zeller (Garcia *et al.* 1994)
This species, which is frequently found in the Colombian llanos, is larger (caterpillar 35 mm) and typically Amazonian, and it seems to be less prolific; its case is straight. The eggs are parasitized by a *Tachinidae* and a *Pteromalidae*. The caterpillars can be affected by a fungus *Beauveria brongniarti* (Sacc.).

Antaeotricha sp. (Genty *et al.* 1978)
The caterpillar, which reaches 18 mm, is yellowish-orange with six wine coloured longitudinal bands. The caterpillar lives between two leaflets bound together with silken threads. Damage is minor, probably because the development of the fungus *Beauveria tenella*. Two other species have been observed ***A. enthruca*** and ***A. phaenora***. They have no economic impact.

COSMOPTERYGIDAE

The species in this family are associated with flowers and do not usually cause serious damage.

Batrachedra arenosella Walker (Nirula 1955, Corbett 1932, Taylor 1930, Lever 1969). Extensive geographical distribution from the West Indies through Central Africa to Australia. Its life cycle lasts around 1 month. Eggs are laid in coconut palm spathes, which are perforated by the caterpillars. Caterpillars can destroy up to 65% of female flowers, though some authors report that they only attack male flowers. In Melanesia, the eggs are parasitized by *Trichogramma* sp. and the caterpillars by *Braconidae* of the genera *Meteorus* sp., *Apanteles* sp. and *Chelonus* sp. The last has been successfully introduced at Flores (Indonesia) (Baringbing 1986).
Two other species have been reported on coconut: ***B. nuciferae*** Hodges, in Venezuela and ***B. perobtusa*** Meyrick (Bondar 1940, in Brazil).

Pyroderces spp., including ***P. amphisarsis*** Meyrick (Wood 1968). Several species of this genus would seem to play a pollination role on oil palm in Malaysia (Tran Yap Pan and Mohd Basri Wahid, unpublished). Following the introduction of *Elaeidobius kamerunicus*, that role is likely to be considerably reduced.

Homoledra sabalella Chambers (Rao 1964).

GELECHIIDAE

Coconympha iriarcha Meyrick observed on coconut fronds on the Malabar coast (India) (Lepesme 1947).

BLASTOBASIDAE

Two species are associated with oil palm flowers: ***Blastobasis byrsodepta*** Meyrick, in central and southern Africa (Lepesme 1947) and ***B. chloroptris*** Meyrick in Malaysia (Wood 1968).

XYLORICTIDAE

Odites sp. on oil palm leaves in Malaysia. The caterpillar lives sheltered in a tube formed by folding over and joining the two edges of a leaflet, like the *Hesperiidae* (Wood 1968).

On the Solomon islands, an unknown species mines a gallery in the petioles of coconut fronds, which are then snapped by the wind (up to 15% of fronds attacked). The eggs are laid on young fronds (19 eggs per female) (Mac Quillan 1969).

OECOPHORIDAE

This family only includes a few species on oil palm and coconut, including the most serious coconut pest in India and Sri Lanka: ***Opisina arenosella*** Walker, which was long known as *Nephantis serinopa* Meyrick. (Ramachandran *et al.* 1979, Child 1974).

Opisina arenosella

Description
The hind wings of the adult, which have a span of 20 to 29 mm, are pale grey speckled with black. The caterpillars are pink with a black head and dark thoracic plates.
Biology
A female lays 140 eggs on average, on the apical section of old fronds. Incubation lasts 3 to 5 days. The caterpillar, which passes through 5 larval instars lasting a total of 5 to 8 weeks, lives under a network of silky threads. Pupation takes 9 to 14 days (Lever 1969). A pheromone emitted by females has been discovered (Murthy *et al.* 1995). Climatic factors probably play a major role in population dynamics in Sri Lanka (Perera *et al.* 1988). The population dynamics of this pest and its parasitoids have been studied (Nadarajan and Channa Basavanna 1980a).
Natural enemies (Dharmaraju 1962, Child 1974, Cock and Perera 1987, Nadarajan and Channa Basavanna 1978).

Parasitoids on caterpillars and chrysalises (Sujatha and Singh 1999).

O. arenosella has a large number of parasitoids in a given region (table 1). For instance, in the state of Karnataka, 5 larva parasites have been inventoried, 13 pupae parasitoids and 11 predators (Sujatha and Singh 1999). However, many of them are rare and the efficacy of numerous others is countered by many hyperparasitoids.

Brachymeria spp. (*Chalcididae)* have a cycle of 12 to 20 days (Pillai and Nair 1993). *B. nosatoi* Habu parasitizes up to 30% of individuals. Also worth mentioning are *B. nephantidis* Gahan and *B. lasus* Walker (Ragwani *et al.* 1997).

Bracon hebetor Say (*Braconidae)* has a cycle of 8-9 days. A female can lay 229 eggs in 14 to 32 hosts with a life span of 20 to 65 days (Sathiamma *et al.* 1986).

Bracon brevicornis Westmael was unsuccessfully released in Sri Lanka from India. Fecundity depends on adult feeding (Yadav 1985). The cocoons are parasitized by *Pediobius brunchicida* (Ronadami) and *P. imbreus* (Walker) (Kapadia 1983).

Trichospilus pupivora Ferrière (*Eulophidae).* A chrysalis endoparasite, with a life span of 16 to 19 days. A female only parasitizes 1 to 2 chrysalises, from which up to 133 adults can emerge.

Antrocephalus hakonensis Ashmead (*Chalcididae),* previously known as *Stomatocerus sulcatiscutellum* (Girault). A chrysalis parasitoid without any great impact. Its biological characteristics vary considerably depending on the temperature (Mohandas and Abduzahiman 1994). It has been successfully released in Sri Lanka.

Elasmus nephantidis Rohwer (*Elasmidae).* A prepupa parasitoid (5 to 10% parasitism). It was released in Sri Lanka but was not found again (Abdurahiman *et al.* 1981).

Goniozus nephantidis (Muesebeck) (*Bethylidae).* Larva parasitoid. The cycle lasts 8 to 14 days. 8 to 26% parasitism seen (Anthony and Kurian 1961). The cocoons are parasitized by the same species as those attacking *Bracon brevicornis.* This parasitoid is itself parasitized by *Aphanogmus goniozi* Dessart (1988).

Apanteles taragmae Viereck *(Braconidae).* This is the main parasitoid of the first and second instars. The larval life cycle lasts 7 days for a total cycle of 18 days. It has numerous hyperparasitoids (table 1), including *Pediobius imbreus (*Walker) (Ghosh and Abdurahiman 1988, Ghosh *et al.* 1993, Pushpalatha and Veeresh 1995).

Eriborus trochanteratus (Morley) *(Ichneumonidae).* 16-day cycle. An adult can parasitize up to 120 caterpillars, i.e. 54% of populations. It is found in Sri Lanka and has been released in India. It has several hyperparasitoids (table 1).

Goryphus sp. *(Ichneumonidae)* can parasitize 60% of chrysalises. It has a life span of 12 to 16 days. *Brachycoryphus nursei* (Cameron), which is usually a boll worm parasitoid, is occasionally seen on *O. arenosella* prepupa and pupa (Pillai and Nair 1987).

Meteoridea hutsoni Nixon *(Braconidae)*. Cycle lasting 17 to 23 days. Parasitoid of larval instars 3 to 5.
Tetrastichus israeli (Mani and Kurian) *(Eulophidae)*. Originating from India and successfully released in Sri Lanka. Numerous studies of this parasitoid have been undertaken with a view to releasing it at different sites in India: effect of the host on development (Nadarayan and Jayaraj 1975), effect of temperature and humidity (Nadarayan and Channabassavana 1978) and the effect of superparasitism (Nadarayan and Jayaraj 1975).
Stomatomyia bezziana Baranoff (*Tachinidae*). This tachinid lays its eggs on old larvae, with a fecundity of 61 eggs. The cycle durations are 5 days for the larva and 9 days for the nymph (Rao and Sudha Rao 1964, Kurian and Sathiamma 1970). Highly parasitized by *Brachymeria nephantidis, Eurytoma albotibialis, Neasteropaeus* sp., *Exoristobia philippinensis* Ashmead, *Pediobius imbreus* (Walker) and *Trichospilus pupivoraz* (Dharmaraju 1962).

Biological control
Numerous attempts have been made to release parasitoids between India and Sri Lanka, usually without success. It was thus that *Bessa remota* (Aldrich), Diptera *Tachinidae*, a parasitoid particularly on *Artona catoxantha* Hampson (*Zygaenidae)* in Southeast Asia, was introduced into India and Sri Lanka. The cycle lasts 18 days and fecundity is 25 to 100 eggs. It has not adapted to *O. arenosella*, but it has to other hosts (Jayanth and Nagarkatti 1984, Sathiamma *et al.* 1987). Likewise, *Trichogramma brasilensis* (Ashmead) was unsuccessfully released in Sri Lanka, like *T. minutum* Riley, which was also released in India. An integrated control programme was proposed by Nadarajan and Channa Bassavana (1980b).
Some coconut varieties seem to be more or less susceptible to attacks by the caterpillar (Kapadia 1981).
Control by male sterilization has been considered (Nadarajan and Channa Bassavana 1980c).
Major reductions in larva populations have been obtained by systemic application of an insecticide containing azadirachtin (Shivashankar *et al.* 2000).

Predators
– Coleoptera *Anthocoridae*: *Triphleps* sp. on eggs, *Cardiastethus exigruis* Poppius and *Carabidae: Parena nigrolineata* (Chaudoir), *Phlaedromius nigrolineatus* (Chaudoir) (Pillai and Kesava Bhat 1987), *Calleida splendida* (F.), *Parena laticincta* Bates (Ganeswara Rao 1978).

– Neuroptera *Chrysopidae*: *Ankylopteryx octopunctata candida* Fabricius is a predator of eggs and caterpillars. (Sathiamma *et al.* 1985).

– Hymenoptera *Formicidae*. Some species are predators of eggs, such as *Technomyrmex albipes* (F. Smith) and *Monomorium* spp., but *Oecophylla smaragdina* (F.) does not seem to have any effect (Way and Cammell 1989)
– Arachnids. Numerous species are found in the crowns. The main ones belong to the genera *Cheiranthium (Clubionidae)* and *Marpissa (Salticidae*) (Sathiamma *et al.* 1987).
– Mites. According to Mathen *et al.* (1968), the mite *Pyemotes ventricosus* Newport appears to be a worthwhile predator which does not seem to have been subsequently used.

Diseases

A polyhedrosis has been observed in the state of Kerala (Philip *et al.* 1982). Other organelles have been isolated, but only from reared insects, as have a bacterium (*Seratia marcescens* Bizio) and a fungus (*Paecilomyces farinosus* Brown and Smith).

Peleopoda spp.

Several species have been observed in different Latin American countries. ***P. arcanella*** Busck in Colombia, Venezuela, Central America. Given its small size (20 mm) the caterpillar causes limited damage, except in the event of severe outbreaks. The cycle lasts 75 days. The parasite complex mainly consists of a Hymenoptera *Encyrtidae* (Genty *et al.* 1978).

In Peru, there are frequent outbreaks of a species of the same genus, despite the existence of a substantial parasite complex: a Hymenoptera *Mimaridae* of the genus *Stethynium* on eggs*; Apanteles* and *Hormius* sp. and more recently *Carinichremylus peleopodae* Achterberg (2000) (*Braconidae), Goniozus* sp. (*Bethylidae*) et *Lymeon* sp. *(Ichneumonidae*) on caterpillars; no fewer than 6 species of Hymenoptera on chrysalises: a *Eulophidae Palmistichus elaeisis* Delvare and Lasalle and five *Chalcididae*: *Brachymeria subconica* Boucek, *B. mnestor* (Walker), *Brachymeria* sp., *Conura celsa* (Walker) and *C. immaculata* (Cresson). Sometimes there is also considerable predatory action: a Hymenoptera *Syrphidae* (Palmas del Espino, Peru unpublished).

Struthocelis semiotarsa (Meyrick) (Genty *et al.* 1978).

This species has been reported in several Latin American countries. The development cycle lasts around 3 months. Damage is slight, causing very typical yellowing of the foliage. Chrysalises are parasitized by the *Eulophidae, Palmistichus elaeisis.*

Lastly, we would point out the species ***Durrantia*** sp. (Ardila y Calvache 1998) in Colombia and ***Machimia flava*** in Ecuadorian Amazonia (Genty, personal communication).

Table 1. Parasitoids and hyperparasitoids of *Opisina arenosella.*

Genera – species	Family	Stage	Location	Hyper parasitism -Reference
Anastatoi brachartonae Gahan	*Eupelmidae*	Pupa	India (very rare)	
Antrocephalus hakonensis	*Chalcididae*	Pupa		Narendran 1985
(Ashmead)	*Chalcididae*	Pupa		
A. cariniceps (Cameron)	*Chalcididae*	Pupa		
A. phaeospilus Waterston	*Chalcididae*	Pupa		
Apanteles taragmae Vierek	*Braconidae*	Caterpillar	Sri Lanka, India	*Aphanogmus manilae* (Ashmead) *Ceraphronidae*, *Eurytoma albotibialis,* *Pediobius imbreus, E. bracanodis,* *Perilampus microgastris,, Eurytoma* sp., *Meteoridea hutsoni* (Nixon)
Dioctes sp.	*Ichneumonidae*	Caterpillar	Sri Lanka	
Bracon brevicornis Wesmael	*Braconidae*	Caterpillar	India	*P. Imbreus, Pediobius* sp., *Elasmus* *nephantidis* (Rohwer)
Goniozus nephantidis Muesebeck	*Bethylidae*	Caterpillar	Sri Lanka, India	*Aphanogmus goniozi* Dessart, *P. imbreus* (Walker), *P. detrimentosus*, *Apha manilae* *Pyemotes ventricosus*
Bracon serinopae (Cherian)	*Braconidae*	Caterpillar	Sri Lanka, India	*Pediobius* sp. *E. albotibialis,* *P. brunchicida (Ronadami)*
Brachymeria exacarinata Gahan	*Chalcididae*			
B. hime atteviae (Joseph,	*Chalcididae*	Pupa	India (very rare)	
Narendran, Joly)		Pupa	India (very rare)	
B. lasus (Walker)	*Chalcididae*			
B. nephantidis Gahan	*Chalcididae*	Pupa	India	
B. nosatoi Habu	*Chalcididae*	Pupa	Sri Lanka, India	*Eurytoma* sp.
Elasmus nephantidis Rohwer	*Elasmidae*	Pupa	Sri Lanka, India	Very important in Kerala
Eriborus trochanteratus (Morley)	*Ichneumonidae*	Pupa	India	*Pleurotropis* sp.
		Prepupa	Sri Lanka	*A. manilae, E. albotibialis*, *B. nephantidis* *P. imbreus, G. nephantidis*

Table 1 (following). Parasitoids and hyperparasitoids of *Opisina arenosella.*

Eurytoma sp.	*Eurytomidae*	Caterpillar	Sri Lanka	
Eurytoma albotibialis Ashmead	*Eurytomidae*	Pupa	Sri Lanka	
Fornicia sp.	*Braconidae*	Caterpillar	Sri Lanka	
Brachycoryphus (Goryphus) nursei (Cameron)	*Ichneumonidae*	Pupa	India	Ectoparasitoid of prepupa and pupa. Cycle of 14 à 17 days. Of little importance. Pillai, Nair 1987a.
Megaselia sp.	*Phoridae*	Pupa	India (very rare)	
Meteoridea hutsoni (Nixon)	*Braconidae*	Caterpillar	Sri Lanka, India	
Pyemotes ventricosus (Newport)	*Acaria*	Caterpillar	Sri Lanka, India	
Stomatomyia bezziana Baranoff	*Tachinidae*	Caterpillar	Sri Lanka	
				B. nephantidis, Eurytoma albotibialis, Exoristobia philippinensis Ashmead - *Neasteropaeus* sp. *(Encyrtidae)* sp., *P. imbreus, Trichospilus pupivora, Syntomosphyrum obscuripes* Ferr. *(Eulophidae)*
Tetrastichus israeli (Mani et Kurian)	*Eulophidae*			
Thelairodrino gracilis (Mesnil)	*Tachinidae*	Caterpillar	India	
Trichospilus pupivora Ferrière	*Eulophidae*	Pupa	India	
Xanthopimpla nana Schulz	*Ichneumonidae*	Pupa	Sri Lanka, India	
X. punctata (F.)	*Ichneumonidae*	Pupa	Sri Lanka	
Xanthopimpla sp.	*Ichneumonidae*	Pupa	Sri Lanka	Pillai, Nair 1987a: egg to adult 12 to 20 days
Thratala sp.	*Ichneumonidae*	Pupa	Sri Lanka	
Winthemia sp.	*Ichneumonidae*	Pupa	Sri Lanka	
Dioctes sp.	*Ichneumonidae*	Pupa	India	
		Caterpillar	Sri Lanka	*Ceraphron* sp. *(Ceraphronidae) E. albotibialis, G. nephantidis, B. nephantidis*

COSSOIDEA

Only two species have been observed in this superfamily, belonging exclusively to the family of the *Cossidae.*

COSSIDAE

Acritocera negligens (Simmonds 1924, 1951; Taylor 1930).
Species located in Fiji. The caterpillar bores into the closed spathe of the coconut inflorescence, but only attacks male flowers. The development cycle lasts 3 months, including 6 weeks for pupation, which takes place in the soil.

Pseudocossus uliginosus (Kerrich)
Madagascan species, whose caterpillar bores into the leaf rachies of the oil palm, mainly in the apical section, which limits the damage caused. The leaf breaks at the gallery point, then dries out. Several fronds per palm can be attacked in this way (Mariau, unpublished).

SESIOIDEA

In this superfamily, only one very serious oil palm and coconut pest is known; it is widespread throughout the countries of South America.

CASTNIIDAE

Castnia daedalus Cramer (Bodkin 1913, Dinther 1956, Anon. 1959, Korytkowski and Ruiz 1980, S. Schuiling and Dinther 1980).

Description
The female, which is slightly larger than the male, can have a wingspan exceeding 20 cm. It is generally greyish-brown in colour, with a green and purple metallic sheen. The flattened caterpillar is highly segmented and has the face of a cerambycid larva. At the end of its development, it can reach 13 cm. The brown chrysalis can be up to 10 cm long; the abdominal segments have two rows of spines.

Biology
Flight, which occurs at nightfall, greatly resembles that of a bat. The female lays its eggs (300) in oil palm crowns. When the eggs, laid on a bunch, hatch, the caterpillars enter the bunch, where part of their development takes place and latter in the peripheral part of the stem. At the end of their development, the caterpillars move towards the surface and pupation takes place in a fibrous cocoon. On coconut, the entire development takes place in the stem.

The cycle lasts around a year (eggs: 2 to 3 weeks; caterpillar: 8 to 12 months; pupation: 30 to 40 days). The eggs are parasitized by a Hymenoptera, *Encyrtidae* of the genus *Ooencyrtus* af. *calpodicus* (Ruiz and Korytkowski 1979) and are attacked by several species of ants of the genera *Odontomachus*, *Pheidole*, and *Iridomyrmex*. A *Nodavirus* type virus has been identified, but does not seem to have any significant impact in regulating populations (Palmas del Espino and Zeddam, personal communication).

Control methods
Biological control trials with nematodes have not given positive results (Segeren 1984). Attacked plantings require strict phytosanitary monitoring (Mariau and Huguenot 1983). Control primarily relies on cultural techniques. Harvesting all bunches and regular pruning of old fronds removes a substantial proportion of the population, but it can prove necessary to carry out chemical control (Huguenot and Vera 1981).

C. licus Drury. Smaller than ***C. daedalus***, this species seems to be located more in central and southern America (Wolcott 1933). No outbreaks of this species have been reported.

ZYGAENOIDEA

ZYGAENIDAE

This family only contains three species living off palms, but all three are or have been serious pests.

Artona catoxantha Hampson (Tothill *et al.* 1930, Van der Vecht 1950, Lever 1953 and 1964, Ooi 1978, Peter 1987).
Reported in Malaysia, Indonesia, PNG, Philippines.

Description
The adult has a wingspan of around 15 mm. The dorsal side of the wings is brown edged with yellow and the ventral side uniformly yellow. The caterpillar is pale in colour, with a purple median line and lighter red lateral lines. The cocoon is elliptical and whitish.

Biology
The complete cycle (5 larval instars) lasts 30 to 40 days. In some countries, such as Malaysia, *A. catoxantha* is the most serious pest on coconut. Between 1970 and 1977, 16 outbreaks were recorded in that country. They mainly occur in the dry season and disappear during very heavy rainfall. The caterpillar browses on the underside of fronds,

leaving the upper epidermis intact during the early instars. Numerous parasitoids have been observed:
- *Tachinidae*
Bessa (Ithychomyia) remota Ald. is the main parasitoid in Malaysia. The efficiency of this tachinid is no doubt linked to the absence or presence of numerous hyperparasitoids: the *Chalcididae, Brachymeria punctiventris* (Cameron), *B. lugubris* Walker (also a parasitoid of *A. catoxantha*), *B. apicicornis* Cameron, *B. euploeae* West., the *Eulophidae Pediobius elasmi* (Ashmead) (also a parasitoid of *A. catoxantha*), *P. parvulus* Ferr., *P. detrimentosus* Gahan, *P. ptychomyiae (*Ferrière*)*, *Nesolynx zygaenarum* (Ferrière), *Tachinobia repanda* Boucek, the *Tetrastichidae Syntomosphyrum obscuripes* Ferrière and *Mellitobia* sp. nr. *hawaïensis* Ferrière, *Euplectromorpha viridiceps* Ferrière, the *Eurytomidae Eurytoma albotibialis* Ashmead (These hyperparasitoids were partly inventoried during trials to introduce the tachinid in India against *Nephantis serinopa*).
Cadurcia leefmansi Bac of little importance in Malaysia, but important in Java, is hyperparasitized by *Syntomosphyrum.obscuripes.*
- *Braconidae*
Microgaster sp. and *Apanteles* sp. in the Philippines.
Apanteles artonae Rohw on second instar caterpillars is itself parasitized by the *Ichneumonidae Microtoridea* sp. and the *Eurytomidae, Eurytoma albotibialis* Ashmead. Although it can be a primary parasitoid of *A. catoxantha*, the latter is probably responsible for outbreaks of the pest in Java.
A. parasae Rohw.
- *Ichneumonidae*
Goryphus maculiceps and *G. maculipennis* Cameron (Malaysia), *G. javanicus* Roman, *G. fasciatipennis, G. inferus* and *G. bituberculatus* Szepl., and *G. rufobasalis* Betr. in Java.
Mesostenus sp. in Malaysia and Java.
- *Eurytomidae: Neoplectrus bicarinatus* Ferrière, ectoparasitoid of *N. maculatus* Ferrière.
- *Eupelmidae*: *Eupelmus catoxanthae* Ferrière also a hyperparasitoid, *E. javae* Gir.
- *Eulophidae: Euplectromorpha artonae* Ferrière.
The *Carabidae Callimerus arcufer* Chapin is a predator but also attacks *B. remota*.

Levuana iridescens Bethune- Baker (Tothill *et al.* 1930). Species localized in Fiji. Biology and damage very similar to those of ***A. catoxantha***. Prior to the introduction of the tachinid *Bessa remota* in 1925, *L. iridescens* was the major coconut pest on this island. The populations have been very well controlled since then.

Homophylotis (Chalconicles) catori (Jordan) (Cachan 1959, Mariau *et al.* 1981).

Description
The adult is 12 mm long. The upper side of the wings is dark brown with a blue sheen, those of the male have a yellow patch on the underside. The caterpillar is golden yellow with purple dorsal lines.

Biology
The cycle lasts 45 to 60 days. The chrysalis is parasitized by several Hymenoptera and a tachinid Diptera, and by an undetermined fungus, which can be highly effective. Serious outbreaks, especially on coconut and sometimes on oil palm, are rare. However, large outbreaks can frequently be seen, but always restricted to the same site.

MEGALOPYGIDAE

All the species belonging to this family, including some very serious pests, are found on oil palm in South America (Genty *et al.* 1978).

Megalopyge albicollis Walker. The caterpillar is 45 mm long and covered with long, dark beige stinging hairs, giving it a cottonwool appearance, with the rear tip culminating in a narrow tuft. The development cycle lasts 102 to 115 days. The caterpillars are gregarious up to the 5th instar. A caterpillar can consume almost 2 leaflets. Parasitoids belonging to the families of the Diptera *Tachinidae* and the Hymenoptera *Chalcididae* and *Ichneumonidae*. ***M. lanata*** Stoll is rarer.

Podalia bolivari Dyar. Caterpillar measuring 20 mm in length, covered with long reddish hairs.

Mesocia pusilla Stoll. The body of the caterpillar, which can reach 30 mm in length, is covered with long, white hairs, giving it a cottonwool appearance. The eggs are laid in parallel rows and are also covered with the same hairs. Pupation takes place in the soil. The biological cycle lasts 69 days. Old caterpillars are attacked by a Diptera *Tachinidae*. Outbreaks remain rare, and that no doubt has something to do with the development, in Colombia, of a very aggressive viral disease (Genty, personal communication).
One other species has been reported: ***M. dumilla*** Dyar.

Norape aff. ***camela***. The adult, which is almost uniformly white, has a wingspan of 15 mm. The 9th instar caterpillar measures 12 mm. It is beige and has long white bristles. The cycle lasts 3 months. Fecundity can be high, up to 400 eggs. Pupation takes place in the soil. The sudden disappearance of large populations in Peru suggests that the caterpillars may be affected by a viral disease (Mariau, unpublished).

N. argyrrhorea Hübner. Has been reported in Peru. The adult is white with red antennae. The caterpillar, which bears long hairs, is speckled with brick red spots. Parasitism consists of a *Eulophidae* on eggs: *Tachinobia repanda* Boucek, on 2nd and 3rd instar caterpillars; a *Tetrastichus* plus two other *Eulophidae* on 4th and 5th instar caterpillars, *Aleiodes* sp. (*Braconidae*) and two *Tachinidae* on 7th and 8th instar caterpillars (Palmas del Espino, unpublished). In very humid conditions, damage by *Norape* is propitious to development of the fungus *Pestalotiopsis,* which can cause more serious defoliation than the caterpillars themselves.

DALCERIDAE

This family only includes one pest, in South America.

Acraga ochracea Walker (Genty *et al.* 1978). The adult, which has a wingspan of 32 mm, is generally yellow ochre in colour. The caterpillar, which reaches 14 mm at the end of its development, has a gelatinous appearance. Its body is covered in translucent, colourless tubercles with brown and yellow designs; the body then turns yellow to milky white.
The cycle is around 2.5 months. As it moves, the caterpillar leaves a typical shiny trail. The chrysalis is protected by a double network of silk.
Attacks occur in the lower crown and always remain slight, probably due to substantial parasitism by the Hymenoptera *Chalcididae* and *Braconidae* and a Diptera *Tachinidae.*

LIMACODIDAE

The family of the *Limacodidae* is widespread throughout the Tropics and palms are the main host of the species in this family which, among the Lepidoptera, is by far the most frequently represented on oil palm and coconut. Indeed, 69 species have been inventoried in Asia, mainly in the Southeast, 20 in tropical America and only 6 on the African continent (Mariau 1999). Coconut, which originated from the Asia-Pacific zone, is probably the original host of a certain number of species in the genera ***Parasa, Thosea, Setora*** and ***Penthocrates.*** Numerous species have adapted very well to oil palm in both Southeast Asia (***Setothosea asigna*** Van Eecke, ***Darna bradleyi*** Holloway for example, Mariau *et al.* 1991) and in tropical America (***Euprosterna eleasa*** Dyar, ***Sibine*** spp. etc (Genty *et al.* 1978). *Limacodidae* are frequently highly polyphagous; for instance ***Setora nitens*** Walker has been inventoried on 32 different genera of plants (Cock *et al.* 1987). Lastly, for many species, notably in Southeast Asia, the host is unknown. The latter are therefore not included in this inventory, but in view of the low population levels with which they are usually associated, it is likely that some of them may feed off palms and have so far gone unnoticed. Table 2 lists all the species found on oil palm and coconut, but only 32 of the 95 species have given typical outbreaks. The biology of the 63 species found in small populations, or as isolated individuals, is obviously much less well known.

Description and biology

Only the characteristics and mean data relative to the species of the family will be indicated here; more precise information can be found in the literature (Genty *et al.* 1978, Mariau et *al.*1981, Mariau *et al.* 1991, Cock *et al.* 1987 in particular).

The eggs are flat and slightly elongated, sometimes looking like those of certain scale insect species. They are laid on the underside of leaflets, sometimes individually (***Euprosterna eleasa*** for example), but usually in groups of more than 100 (***Sibine*** spp.). The situation corresponds to the caterpillars, which are either solitary or gregarious. Incubation lasts 4 to 8 days on average.

The caterpillars are very typical: brightly coloured, often richly decorated, and their bodies bristle with protuberations and numerous highly irritant hairs, hence their being known as needle caterpillars. The designs and layout of the hairs varies depending on the instars, which range from 7 to 9, with a cycle of 30 to 40 days. A few species, such as those belonging to the genera ***Olona*** or ***Cheromettia***, are hairless and completely smooth (gelatine caterpillar). The body is legless and holds on to the substrate by adhesion. Movement, which is also highly typical, is similar to that of slugs, with an ondulating contraction from front to rear, which is why caterpillars in the *Limacodidae* family are also sometimes called slug caterpillars.

The hard, silken cocoons are ovoid in shape and sometimes slightly elongated, measuring from 15 to 20 mm. They are usually brown, sometimes a dirty white. They are also covered in stinging hairs and, in some cases, covered with a web, like those of ***Latoia pallida*** Möschl. Pupation lasts 30 to 40 days. and often takes place on fronds or at the base of the rachis. However, in some species, the caterpillar drops to the soil, where it constructs its cocoon (***Setora nitens*** in Asia, ***Euprosterna eleasa*** in America, for example).

The adults have a wingspan varying from 20 to 40 mm, though a few small species, such as those of the genus ***Penthocrates***, do not exceed 5 to 6 mm. The adults are generally brown, sometimes blackish-brown, or straw yellow. Adult activity is usually nocturnal and fecundity is high: several hundred to almost a thousand eggs.

The meeting of the sexes can be via pheromone substances, as in ***Setothosea asigna*** (Zagatti *et al.* 1993, Sasaerila *et al.* 1997), two species of ***Darna*** (Sasaerila *et al.* 2000a) and ***Setora nitens*** (Sasaerila *et al.* 2000b).

Natural enemies The following three groups will be examined in turn: predatory insects (predation by birds and other vertebrates is marginal), parasitoid insects, and pathogens (Cock *et al.* 1987, Sipayung *et al.* 1989).

Predators

Apart from the predatory activities of some carabid Coleoptera, the main predators are Hemiptera belonging to the families *Pentatomidae* and *Reduviidae*. These bugs are highly polyphagous and can feed off virtually any caterpillar; the hosts reported in the literature are therefore highly specific in nature. The species observed primarily belong to two families:

– *Pentatomidae: Chanteconidea furcellata* (Wolff) (Ghorpade 1976, Desmier de Chenon *et al.* 1989) is the most common species, *Chanteconidea* sp. has been reported in Sumatra on ***Setora nitens*** (Kalshoven and Van der Laan 1981), *Alcorrhynchus grandis* (Dallas) in South America (Genty 1972) on ***Sibine*** spp. in particular, along with *Podissus agenor.*

– *Reduviidae: Cosmolestes picticeps* and *Isyndus heros* Stål on ***Setothosea asigna*** (Tiong 1982), *Sycanus* sp., *S. dichotomus* (Stål), *S. leucomesus* Walker and *S. macracanthus* Stål (Tiong 1982, Kalshoven and Van der Laan 1981) on various Asian species.

– *Pyrrhocoridae: Dindymus rubiginosus* (Fabricius) is very rare on ***Setothosea asigna*** (Gonggrijp 1931).

Parasitoids

These belong to the orders Hymenoptera and Diptera. The parasite complexes are given in table 2. Only the main parasitoids are mentioned here.

Hymenoptera

These belong to numerous families, and to four superfamilies:

Ichneumonoidea

– *Ichneumonidae, Cryptinae.* Parasitism usually occurs on the caterpillar in the final instar, with adults emerging from the chrysalis. *Buysmania oxymora* (Tosquinet) has been observed on 11 different hosts. *Chlorocryptus purpuratus* (Smith) has been found on 9 species of *Limacodidae.*

– *Braconidae.* In the parasite complex of most species of *Limacodidae*, there is at least one species belonging to this large family. Four main genera are identified: *Apanteles, Cotesia, Spinaria* and *Fornicia.* In the genus *Apanteles*, 6 species have been determined on limacodids in Southeast Asia. The main ones are *A. aluella* Nixon on at least 8 hosts mainly belonging to the genera ***Darna*** and ***Thosea***. On ***Parasa***, it is *Apanteles parasae* Rhower that is found, whereas ***Penthocrates*** are parasitized by *A. penthocratus* Austin. In the genus *Spinaria*, it is *S. spinator* (Guérin) that usually attacks species of the genus ***Setora***. Among the species of the genus *Fornicia*, it is *F. ceylonica* Wilkinson that is the most common, notably on caterpillars of the genus ***Darna***. Five or six other species are rarer and usually monospecific, such as *F. tagalog* (Cushman) on ***Susica malayana*** Hering. In tropical America, *Braconidae* belonging to the genera *Apanteles, Fornicia* and *Cotesia* are very often found on the majority of *Limacodidae* caterpillars.

Chalcidoidea. This superfamily includes a very large number of parasitoid or hyperparasitoid species that belong to four families of unequal importance: *Chalcididae, Eurytomidae, Eulophidae* and *Trichogrammatidae.*

– *Chalcididae.* In the genus *Brachymeria*, a single species has been identified in Asia: *B. lasus* (Walker) (Cock *et al.* 1987). It is always a hyperparasitoid of the *Braconidae, Ichneumonidae* and *Tachinidae.* Another species has been obtained from ***Parasa lepida*** in India (parasitoid or hyperparasitoid?). In tropical America, several species are parasitoids of various other leaf-eating Lepidoptera on oil palm, along with Coleoptera *Hispinae*, but on *Limacodidae*, 4 species of the genus *Conura* are hyperparasitoids through *Casinaria* sp. (*Ichneumonidae)* and *Cotesia* sp. *(Braconidae*) (Delvare 1993). Four other species are also primary parasitoids: *Conura (Conura) papo* Delvare, on an undetermined *Limacodidae* in Ecuador, *C. immaculata,* on ***Sibine fusca***, and lastly *C. martinezi* and *C. phobetronae* on ***Phobetron*** sp.

– *Eurytomidae* and *Eupelmidae.* The 4 species listed on Asian *Limacodidae* are hyperparasitoids through *Braconidae* and *Tachinidae.*

– *Eulophidae.* In this family, on Asian *Limacodidae* caterpillars, around thirty species have been observed that are very often primary parasitoids (Cock, 1987) (table 2). The main species belong to the genera *Aroplectrus, Euplectromorpha, Neoplectrus, Platyplectrus* and *Nesolynx*, the last genus living more particularly off the genus ***Penthocrates***. A few species of *Eulophidae* are hyperparasitoids on *Braconidae* of the genus *Apanteles* and on *Tachinidae.* In tropical America, little is known about the fauna of the *Eulophidae.* It includes a recently described species: *Palmistichus elaeisis* Delvare and LaSalle, a hyperparasitoid on ***Sibine fusca*** through *Cotesia* sp. and *Elasmus maculatus* Howard. On the other hand, *P. elaeisis* is a primary parasitoid on numerous species of other families of Lepidoptera and Coleoptera pests of the oil palm (Delvare and LaSalle, 1993).

– *Trichogrammatidae.* Generally speaking, there is little parasitism on the eggs of *Limacodidae.* In Asia, *Trichogrammatoidea thoseae* Nagaraja has been observed on 4 species. In America, no *Trichogrammatidae* have been reported on *Limacodidae.*

Bethyloidea and Ceraphronoidea. The following families are attached to these two superfamilies:

– *Chrysididae.* Two species of this family are parasitoids on *Limacodidae* in Asia: *Praestochrysis shanghaiensis* (Smith) and another species of the same genus, both notably on ***Parasa lepida*** (Polaszek, 1987).

– *Ceraphronidae*. In Asia, this family only includes two species that are hyperparasitoids on *Limacodidae* through *Braconidae* and *Ichneumonidae: Aphanogmus manilae* (Ashmead) and *Ceraphron* sp. (Austin, 1987).

Diptera

The Diptera, and primarily those of the family *Tachinidae,* play a major role in regulating *Limacodidae* populations. The *Bombyliidae* are much less numerous, and their parasitic activity is therefore less extensive.

– *Tachinidae*. In Asia, the main species *Chaetexorista javana* Brauer and Bergenstamm has been observed on 17 different hosts (Harris, 1987) including major leaf-eaters such as ***Parasa lepida, Setora nitens, Setothosea asigna***. In the same tribe of the *Exoristini*, the following species are worth mentioning: *Austrophorocera grandis* (Macquart) on three hosts living off palms, *Bessa remota* (Aldrich) on two hosts and *Exorista sorbillans* (Wiedemann) on three hosts. In the tribe of the *Sturmiini*, we would mention two species of the genus *Palexorista: P. solennis* (Walker) and *P.* sp. nr. *ophirica* from the Indo-Malaysian peninsula. Lastly, two other species, belonging to other tribes, have only been observed on a single host each: *Nealsomyia rufella* (Bezzi) on ***Chalcocelis albiguttatus*** Snellen and *Compsilura concinnata* (Meigen) on ***Thosea*** sp. (Bezzi, 1925, Harris, 1987). In Africa, a single species on ***Latoia viridissima*** Holland: *Palexorista* sp. nr. *mayneana* (Igbinosa, 1988b). In America too, a single species has been reported, *Palexorista coccyx* on ***Sibine fusca*** (Genty, 1972).

– *Bombyliidae*. All the species observed on the *Limacodidae* of palms belong to the genus *Systropus*. In Asia, *S. polistoides* Westwood has been observed on one species and *S. roepkei* de Meijere on 6 hosts (Greathead, 1987). In Africa, *S. pelopoeus* (?) on ***Latoia viridissima*** (Igbinosa, 1988b), and in America, *Systropus nitidus* Wiedemann on ***Sibine fusca*** (Genty, 1972).

– *Ceratopoginidae.* The adult makes feeding punctures on different species of *Limacodidae* caterpillars in tropical America.

Table 2. Species of ***Limacodidae*** and their parasitoids on oil palm and coconut.

Parasitoids: H.: Hymenoptera – D.: Diptera – Br: *Braconidae* – I.: *Ichneumonidae* – C.: *Chalcididae* – E.: *Eulophidae* – S.: *Sarcophagidae* – Bo.: *Bombyliidae* – T.: *Tachinidae* – Tr.: *Trichogrammatidae*.
Countries: A: Argentina – A: Australia – Al: Andaman Islands – Br: Burma – C: Colombia – CA: Central America – CAf: Central Africa – CI: Côte d'Ivoire – CR: Costa Rica – E: Ecuador – EAf: East Africa – G: Guiana – Ga: Guatemala – Gh: Ghana – H: Honduras – I: Indonesia (Sm: Sumatra – Bl: Bali – Bo: Borneo – J: Java – Sl: Sulawesi – Mo: Moluccas) – In: India – M: Malaysia (Sa: Sabah) – Me: Mexico – P: Peru – Pl: Philippines (Lz: Luzon – Cb: Cebu – Ne: Nepos – Ly: Leyte) – PNG: Papua New Guinea – SA: South America - SEA: South East Asia - T: Thailand – Tr: Trinidad – Ve: Venezuela – Vi: Vietnam – WAf: West Africa
Hosts: P: oil palm – C: coconut.
* **Species in which serious to very serious outbreaks have been observed**.

NAME	COUNTRY	HOSTS	PARASITOIDS	REFERENCES
Baria elsa Druce	CI	P.	?	Mariau, 1982
Birthamoides circinatus Snellen	P I (Sl)	C.	H. *Br.*: *Apanteles aluella* H. *E.*: *Neoplectrus* sp.	Cock *et al.*, 1985 Holloway *et al.*, 1987
Birthamula chara Swinhoe	M I (Sm)	P.	H. *Br.*: *Fornicia ceylonica* H. *Ich.*: *Chlorocryptus purpuratus* H. *E.*: *Euplectrus* D. *T.*: *Chaetexorista javana* D. *Bo.*: *Systropus roepkei*	Wood, 1968 Mariau *et al.*, 1991
Birthosea bisura* Moore	M I (J Sm)	P.	H. *Br.*: *Fornicia* sp. H. *E.*: *Euplectromorpha* spp. H. *I.*: *Buysmania oxymora, Chlorocryptus purpuratus* D. T.: *Chaetexorista javana*	Kalshoven & Ven der Vecht, 1950 Wood, 1968 Mariau *et al.*, 1991 Hertslet et Duckett, 1971
Cania bandura Moore	I (Sl B J Bo) T Br M	C. and P.	H. *Br.*: *Apanteles caniae* + others *Br.* D. *Tach.*: *Corcelia rufiventris*	Holloway, 1986 Wood, 1968 Sudharto *et al.*, 1990
Cania robusta Hering	T M Br	C. and P.	?	Corbett, 1932

NAME	COUNTRY	HOSTS	PARASITOIDS	REFERENCES
Cania siamensis Tams	T	C.	H. *I.*: *Buysmania oxymora* H.: *Chrysididae* H. *E.*: *Platyplectrus*. *T.* ?	Holloway, 1986
Cania striola Hering	M I (Su Bo)	P.	?	Holloway, 1986 Kalshoven & van der Vecht, 1950
Casphalia extranea Walker	C I	C.and P.	?	Fédière *et al.*, 1986 Fédière, 1996
Chalcoscelides castaneipars Moore	Br	C.	?	Kalshoven & van der Vecht, 1950
Chalcocelis albiguttatus Snellen	V T M I (Su Sl) Pl PNG A	C. and P.	D. *Bo.*: *Systropus roepkei* H. *Br.*: *Fornicia chalcoscelidis, Batotheca* sp. H. *I.*: *Buysmania oxymora, Goryphus* sp. H. *C.*: *Brachymeria lasus* (hyperp.) H. *E.*: *Euplectrini* D. *T.*: *Austrophorocera grandis, Chaetexorista javana, Nealsomyia rufella* D. *S.*: *Boettcherisca peregrina, Sarcorohdendorfia antilope, Seniorwhitea krameri, S. orientalis*	Corbett, 1932, 33, 37 Roepke, 1935 Conway and Tay, 1969
Cheromettia sumatrensis Heylaerts	M I (Sm J Bo)	C. and P.	D. *Bo.*: *Systropus roepkei* H. *Br.*: *Apanteles aluella, A. belippae*	Dammerman, 1929 Wood, 1968
Contheyla rotunda Hampson	In	C.	H. *Br.*: *Aleiodes* sp., *Rogas* sp. H. *Chrysidae*: *Chrysis* sp. H. *C.*: *Antrocephalus* sp.	Nirula *et al.*, 1954 Ayyar T.V.R., 1927 Mathewkutty and Subramaniam 1997
Ctenocompa hilda Druce	CAf	P.	D. *T.* ?	Lepesme, 1947

NAME	COUNTRY	HOSTS	PARASITOIDS	REFERENCES
Darna bradleyi* Holloway	M (Sa) I (Bo Sm)	P.	H. *Tr.*: *Trichogrammatoidea thoseae* H. *E.*: *Neoplectrus* nr. *clavatus* H. *Br.*: *Apanteles aluella*	Kimura, 1978 Mariau *et al.*, 1991
Darna catenatus* Snellen	I (Sl)	C.and P.	H. *Br.*: *Apanteles aluella* H. *Chrysidae*: *Chrysis* sp. H. *E.*: *Neoplectrus arcolatus (Nesolynx* sp. hyperp ?) *Platyplectrus orthocraspedae* D. *T.*: *Bessa remota, Chaetexorista javana*	Tams, 1931 Mariau *et al.*, 1991 Ponto & Mo, 1950
Darna diducta Snellen	T M I (Sm Bo J) Pl (Lz)	C. and P.	H. *Br.*: *Apanteles aluella (?) (Aphanogmus manilae, Rogas* sp., hyperp.) H. *E.*: *Neoplectrus clavatus* H. *I.*: *Buysmania oxymora* D. *T.*: *Chaetexorista javana, Palexorista solennis*	Wood, 1968 Kalshoven & Van der Laan, 1981 Kalshoven & Van der Vecht, 1950 Mariau *et al.*, 1991
Darna furva* Wileman	T	C. and P.	H. *Br.*: *Apanteles* sp., *Fornicia ceylonica* H. *E.*: *Euplectromorpha* sp. *Platyplectrus* sp. H. *I.*: *Paraphylax varius, Melalophacharops papilionis* D. *T.* ?	Holloway *et al.*, 1987
Darna mindanensis Holloway	Pl (Mi)	C. and P.	H. *Br.*: *Apanteles aluella (Aphanogmus manilae, Eurytome* sp. hyperp.), *Fornicia* sp. H. *E.*: *Aroplectrus dimerus* H. *I.*: *Scenocharops punctata (Aphanogmus manilae, Brachymeria lasus*, hyperp.) D. *T.*: *Chaetexorista javana*	Cock *et al.*, 1985
Darna nararia Moore	I (Sl)	C.	H. *E.*: *Neoplectrus maculatus* H. *I.*: *Goriphus varibalteatus* H. *Br.*: *Fornicia ceylonica (Paraphylax varius* Ich. hyperp.)	Kurian, 1954 Menon & Pandalai, 1958

NAME	COUNTRY	HOSTS	PARASITOIDS	REFERENCES
Darna pallivitta Moore	T M I (Bo J)	C. and P.	?	Kalshoven & Van der Vecht, 1950
Darma sordida Snellen*	T M I (Sm J)	C. and P.	H. *Br.: Rogas sp.* H. *I.: Buysmania oxymora*	Piepers & Snellen, 1900
Darna trima * Moore	M I (Bo Sm J)	C. and P.	H. *Br.: Apanteles aluella (Aphanogmus manilae* hyperp.), *Apanteles* sp. *Fornicia penang, Rogas* sp. H. *E.: Platyplectrus orthocraspedae* H. *C.: Brachymeria lasus* (hyperp.) H. *I.: Buysmania oxymora* D. *Bo.: Systropus roekpei* D. *T.: Chaetexorista javana*	Tiong & Munroe, 1977 Corbett, 1932 Mariau *et al.*, 1991
Darna tuaranensis Holloway	I (Bo)	C.	H. *Br.: Apanteles aluella*	Holloway *et al.*, 1987
Episibine intensa* Dyar	C G	P.	Common to *Sibine fusca*	Genty *et al.*, 1978
Episibine sibinides* Dyar	P	P.	?	Genty *et al.*, 1978
Euclea cippus Cramer	SA	P.	?	Genty *et al., 1978*
Euclea cuprostriga Dyar	G S C P	P.	?	Genty *et al.*, 1978
Euclea diversa* Druce	Me H CRC E P	P.	Common to *Sibine fusca*	Genty *et al.*, 1978
Euclea norba Druce	CA	P.	?	Genty *et al.*, 1978
Euclea plugma* Sepp	Me Ga CR H C	P.	?	Genty *et al.*, 1978

NAME	COUNTRY	HOSTS	PARASITOIDS	REFERENCES
Euprosterna eleasa* Dyar	SA	P.	H. *I.*: *Casinaria* sp. (*Theronia* sp., *Spilochalcis* sp., *S. nunani*, *Conura immaculata*, *C. biannulata*, *Lymeon*, hyperp.), *Isdromas* sp. H. *Br.*: *Fornicia* sp., *Apanteles* sp., *Rhogas* sp., *Glyptopanteles* H. *E.*: *Stenomesius* sp. H. *C.*: *Brachymeria* sp., *Pseudobrachymeria* sp. H. Tr. on eggs	Genty, 1976 Genty *et al.*, 1978 Delvare, 1992
Idonauton apicalis Walker	T M I (Sm Bo)	C. and P.	?	Holloway *et al.*, 1987
Latoia pallida* Mösche	WAf CAf	C. and P.	Probably similar to *L. viridissima*	Mariau & Julia, 1973 Mariau *et al.*, 1981
Latoia viridissima* Holland	Waf CAf	C. and P.	H. *Br.*: *Apanteles* sp., *Rogas* sp. H. *E.*: *Tetrastichus* spp. nr. *balteatus* H. *C.*: *Brachymeria* sp., *Chrysis spina* H. *I.*: *Cocygodes coccys* D. *T.*: *Palexorista* nr. *mayneana* D. *Bo.*: *Systropus pelopeus*	Igbinosa, 1985, 1988 Allen & Bull, 1984 Mariau & Julia, 1973 Mariau *et al.*, 1981
Microthosea celaena West	Pl (Mi Ly)	C.	?	Cock *et al.*, 1985
Microthosea minima Semper	Pl (Lz Cb Ne)	C.	H. *Br.*: *Fornicia annulipes*, *Spinaria armata* H. *E.*: *Metaplectrus thoseae*, *Euplectromorpha malayensis* D. *T.*: *Austrophorocera grandis*	Banks, 1906 Cock *et al.*, 1985
Microthosea porthetes Tams	I (Sl)	C.	H. *Br.*: *Fornicia thoseae*	Tams, 1931
Narosa rosipuncta Holloway	M I (Sm Bl)	P.	H. *I.*: *Buysmania oxymora* D. *Bo.*: *Systropus roepkei*	Holloway, 1986 Sudharto *et al.*, 1990
Natada fusca Druce	Me CA C P Ve	P.	?	Genty *et al.*, 1978

NAME	COUNTRY	HOSTS	PARASITOIDS	REFERENCES
Natada michorta Dyar	CR H C G	P.	?	Genty *et al.*, 1978
Natada pucara Dognin.	SA	P.	?	Genty *et al.*, 1978
Natada subpectinata * Dyar	Tr S C E	P.	H. *I. Casinaria* sp. *(Conura immaculata* hyperp.) H. *E.*: *Horismenus* spp.	Genty *et al.*, 1978 Delvare, 1992
Olona albistrigella Snellen	I (J)	C.	?	Piepers & Snellen, 1900 Kalshoven & van der Vecht, 1950
Olona gateri West	M I (Sm)	C. and P.	H. *E.*: *Euplectromorpha* sp. H. *Br.*: *Apanteles* sp.	Wood, 1968 Sudharto *et al.*, 1990 Holloway *et al.*, 1981
Parasa sp.	Pl (Mi)	C.	H. *Br.*: *Apanteles parasae (Pediobius elasmi, P. ptychomyiae* ? hyperp.) H. *E.*: *Eurytoma* sp. *(Aphanogmus manilae* hyperp.) H. *I.*: *Buysmania oxymora* H. *C.*: *Brachymeria lasus* (hyperp.) D. *T.*: *Chaetexorista javana, Sarcosolomia crinita*	Aterrado & Abad, 1986
Parasa balitkae Holloway	I (Sl)	C.	?	Davis & Sudasrip, 1982
Parasa chlorostigma Snellen	I (Sl)	C.	?	Holloway *et al.*, 1987
Parasa corbetti Holloway	M	C.	?	Holloway *et al.*, 1987
Parasa darma Moore	I (Sm J) P M T Br	C.	?	Holloway *et al.*, 1987
Parasa darmoides Holloway	Pl	C.	*E.* ?	Holloway *et al.*, 1987

NAME	COUNTRY	HOSTS	PARASITOIDS	REFERENCES
Parasa lepida* Cramer	SEA	C.	*Br.*: *Apanteles parasae (Elasmus albomaculatus* (hyperp.), *Rhogas* sp., *Fornicia ceylonica*, *Eurytoma* sp., *Clinocentrus* sp. *C.*: *Kriechbaumerella ayyari*, *Brachymeria lasus* (hyperp. Ich.) *E.*: *Euderastichus* sp., *Euplectromorpha malayensis*, *Pediobius* sp. (hyperp.) *Eurytomidae: Eurytoma monemae* (hyperp.) *Eupelmidae*: *Eupelmus catoxanthae* (hyperp. T.) *I.*: *Buysmania oxymora, Trachysphyrus oxymorus* *S*: *Sarcophaga antilope, Seniorwhitea orientalis* *Tach.*: *Bessa remota, Tachina munda, Exorista sorbillans, Blepharella lateralis, Chaetexorista javana (Aphareta* sp. hyperp.)	Desmier de Chenon, 1982 Corbett, 1937 Pillai & Ayyar 1968
Parasa philepida* Holloway	Pl (Mi)	C.	*Br.*: *Apanteles parasae (Aphanogmus manilae, Eurytoma* sp., *Pediobius elasmi*, hyper) *Eurytomidae*: *Eurytoma monemae* (hyperp. ?) *I.*: *Buysmania oxymora, Stictopisthus* sp. (hyperp.) *S.*: *Sarcolosomia crinitus* ? *T*: *Chaetexorista javana (Brachymeria lasus, Pediobius ptychomyiae*, hyperp.)	Cock, 1987
Parasa sundalepida Holloway	M I (Sl Bo)	C.	?	Corbett, 1932
Pectinarosa alastor* Tamps	I (Sl)	C.	?	Davis & Sudasrip, 1982 Tams, 1931
Penthocrates albicapitata Holloway	Pl (Lz)	C.	*Br. Apanteles penthocratus* *E. Aroplectrus dimerus, Nesolynx* sp. *I.*: *Prionacis ligaoi*	Cock *et al.*, 1985, 1986 Holloway *et al.*, 1987
Penthocrates bigenita Meyrick	I (J)	C.	*Br.*: *Apanteles* sp. *E.*: *Euplectrus* sp.	Kaslhoven & van der Vecht, 1950

NAME	COUNTRY	HOSTS	PARASITOIDS	REFERENCES
Penthocrates rufa Holloway	Pl (Cb)	C.	*Br.*: *Apanteles penthocratus* *E.*: *Aroplectrus dimerus, Nesolynx* sp. *Eupelmidae*: *Eupelmus* sp.	Cock *et al.*, 1985, 1986 Holloway *et al.*, 1987
Penthocrates rufofascia* Holloway	P (Lz Ne)	C.	*Br.*: *Apanteles penthocratus* *E.*: *Aroplectrus dimerus, Nesolynx* sp. *Eupelmidae*: *Eupelmus* sp. *I.*: *Prionacis ligaoi*	Cock *et al.*, 1985, 1986 Holloway *et al.*, 1987
Penthocrates styx Holloway	Pl (Mi)	C. and P.	*Br.*: *Apanteles penthocratus* *E.*: *Aroplectrus dimerus, Nesolynx* sp.	Cock *et al.*, 1985, 1986
Penthocrates zelaznyi* Holloway	Pl (Mi)	C.	*Br.*: *Apanteles penthocratus* *E.*: *Aroplectrus dimerus, Nesolynx* sp. *I.*: *Prionacis ligaoi*	Cock *et al.*, 1985, 1986 Holloway *et al.*, 1987
Phobetron hipparchia Cramer	SA CA	P.	*Br.*: *Apanteles* spp. *I.* ? *Br.*: *Conura martinezi,* *C. phobetronae* *T.* ?	Genty *et al.*, 1978 Delvare, 1992
Pseudonirmides sola Swinhoe	I (Sl Bo) Pl (Lz Mi)	C.	*Br.*: *Apanteles* sp. (*Pediobius elasmi* (hyperp.) *I.*: *Buysmania oxymora*	Cock *et al.*, 1985 Holloway, 1986
Pygmaeomorpha priori Holloway	PNG	C.	?	Holloway *et al.*, 1987
Pygmaeomorpha ramosa Hering	PNG	C.	?	Hering, 1931
Pygmaeomorpha rubrimixta Hering	PNG	C.	?	Hering, 1931

NAME	COUNTRY	HOSTS	PARASITOIDS	REFERENCES
Quasithosea sythoffi Snellen	Br M I (J)	P.	*Chalcidoidea* ?. *E.*: *Euplectrus* sp. *T.*: ?	Kalshoven & van der Laan, 1981
Setora cupreistriga* Walker	I (Bo)	C. and P.	*Br.*: *Fornicia* sp., *Spinaria spinator* *C.*: *Brachymeria* spp., Par. et hyperp. *Encyrtidae*: *Encyrtus* *Eurytomidae*: *Eurytoma* (hyperp.) *I.*: *Goryphus bunoh*, *Goryphus* sp., *Lienella* sp., *Xanthopimpla* sp. *Tr.*: *Trichogrammatoidea thoseae* *T.*: *Chaetexorista javana*	Conway & Tay, 1969 Syed & Pang, 1972
Setora kelapae* Holloway	I (Sl)	C.	?	Davis & Sudasrip, 1982 Soekarjoto *et al.*, 1980
Setora longifurca Holloway	Pl (Mi Ly)	C.	?	Cock *et al.*, 1985
Setora nitens* Walker	M I (J Sm)	C. and P.	*T.*: *Trichogrammatoidea thoseae* *Br.*: *Fornicia ceylonica*, *Fornicia* sp., *Apanteles parasae*, *Meteorus* sp., *Spinaria spinator*, *Rogas* sp. (*Brachymeria lasus* nr hyperp.) *I.*: *Buysmania oxymora*, *Chlorocryptus purpuratus*, *Goryphus mesoxanthus*, *Theronia orientalis*, *Xanthopimpla* sp. *Eurytomidae*: *Eurytoma* sp. (hyperp.) *T.*: *Chaetexorista javana*, *Exorista sorbillans*, *Palexorista* sp. *S.*: *Seniorwhitea orientalis*	Corbett, 1932 Miller, 1929 Piepers & Snellen, 1900 Wood, 1968 Mariau *et al.*, 1991 Soekarjoto *et al.*, 1980 Lever, 1969
Setora tagalog Holloway	Pl (Lz)	C.	*Br.*: *Triraphis* sp., *Eurytoma* sp. *E.*: *Euplectromorpha malayensis* *T.*: *Chaetexorista javana*	Cock *et al.*, 1985 Holloway *et al.*, 1987

NAME	COUNTRY	HOSTS	PARASITOIDS	REFERENCES
Setothosea asigna* Van Eecke	M I (Bo Sm) Pl	C. and P.	*Br.*: *Rogas* sp., *Spinaria* sp., *Fornicia* sp., *Fornicia ceylonica (Aphanogmus manilae, Ceraphronidae* : hyperp.) *E.*: *Euplectromorpha* nr. *bicarinata, Metaplectrus solitarius* *I.*: *Chlorocryptus purpuratus, Goryphus* sp., *Theronia orientalis* *Tr.*: *Trichogrammatoidea thoseae* *T.*: *Chaetexorista javana*	Tiong, 1979, 1982 Wood & Nesbit, 1969 Kalshoven & Van der Vecht, 1950 Mariau *et al.*, 1991 Kimura, 1978
Sibine apicalis Dyar	P	P.	Common to *Sibine fusca*	Martignoni & Iwai, 1981
Sibine fusca* Stoll	SA	P.	*Br.*: *Apanteles* sp. nr. *glomeratus (Elasmus maculatus, Spilochalcis* sp., hyperp.), *(Palmistichus elaeisis* paras. of *Elasmus*) *I.*: *Casinaria* sp. *(Cotesia* sp., *Palmistichus elaeisis, Conura immaculata, C. camescens, C. biannulata*, hyperp.) *T.*: *Palpexorista coccyx* *Bo.*: *Systropus nitidus*	Genty, 1972 Genty *et al.*, 1978 Delvare, 1992, Delvare & La Salle, 1993
Sibine megasomoides Walker	Ve C E P	P.	Common to *Sibine fusca*	Genty *et al.*, 1978
Sibine modesta Cramer	A G Tr	C.	?	Wolcott, 1933 Lepesme, 1947
Sibine nesea* Stoll	S E C	P.	Common to *Sibine fusca*	Genty *et al.*, 1978
Sibine pallescens Dognin	Ve C	P.	?	Genty *et al.*, 1978
Sibine trimaculata Sepp	A G	C.	?	Lepesme, 1947
Susica malayana Hering	T M I (Sm) Pl	C. and P.	*Br.*: *Fornicia tagalog (Aphanogmus manilae* hyperp.) *I.*: *Chlorocryptus purpuratus* *T.*: *Chaetexorista javana*	Cock *et al.*, 1985 Wood, 1968 Mariau *et al.*, 1991

NAME	COUNTRY	HOSTS	PARASITOIDS	REFERENCES
Thosea andamanica Holloway	IA	C.	*Br.*: *Fornicia* sp. (*Aphanogmus* hyperp.)	Isaac, 1925 Ansari *et al.*, 1992
Thosea austrocaerulea Holloway	Pl (M)	C.	?	Cock *et al.*, 1985
Thosea bipartita Hering	In BrV T	C.	?	Holloway *et al.*, 1987
Thosea boreocaerulea Holloway	Pl (Lz)	C.	*Br.: Fornicia annulipes* *T.: Ceromya rotundicornis (?)*	Cock *et al.*, 1985
Thoseahrysoparala Tamps	Gh	C.	?	Tamps, 1931
Thosea lutea Heylaerts	M	C. et P.	*Br.: Apanteles* sp. *C.: Brachymeria* sp.(hyperp.) *Chrysididae: Chrysis shanghaiensis* *Eurytomidae: Eurytoma* sp. *I.: Chlorocryptus purpuratus (Stictopisthus* sp. Hyperp.)	Kalshoven & Van der Vecht, 1950
Thosea mediostrigata Hering	M I (Sm)	C.	?	Holloway *et al.*, 1987
Thosea monoloncha* Meyrick	I (Sl Mo) PNG	C. and P.	*Br.: Apanteles roepki* *T.: Austrophorocera grandis*	Roepke, 1935 Greve & Ismay, 1983 Mariau *et al.*, 1991
Thosea phillipina Holloway	P	C.	*Br.: Apanteles aluella (Aphanogmus manilae, Eurytoma* sp., *Tetrastichus* sp. hyperp.), *Spinaria armata (Brachymeria lasus, Theronia orientalis, Paraphylax sorianoi,* hyperp.), *Fornicia moronis* *E.: Metaplectrus thoseae* *I.: Scenocharops punctata* *Tr.: Austrophorocera* sp., *Chaetexorista javana* *Tr.: Trichogrammatoidea thoseae*	Cock *et al.*, 1985
Thosea siamica Holloway	T	C. and P.	*C.: Brachymeria* sp. *Tr.: Trichogrammatoidea* sp.	Holloway *et al.*, 1987

NAME	COUNTRY	HOSTS	PARASITOIDS	REFERENCES
Thosea vetusta Walker	M I (Bo Sm J Sl)	P.	*I.: Chlorocryptus purpuratus* *T.: Chaetexorista javana*	Wood, 1968 Holloway, 1986 Mariau *et al.*, 1991
Thosea unifascia Walker	In	C.	*Br.*: ?	Isaac, 1925
Trichogyia semifascia Hampson	B M I (Sm J) Pl (Mi)	P.	*I.: Chlorocryptus purpuratus*	Holloway *et al.*, 1987
Trogocrada deleter Tams	WAf	C.	?	Child, 1974

Diseases

Whilst parasitoids play a decisive role in keeping *Limacodidae* populations at a stable level, pathogens, primarily viruses and more secondarily fungi, can cause epidemics capable of rapidly checking an outbreak.

Viruses (Desmier de Chenon *et al.* 1988, Entwistle 1987, Mariau and Desmier de Chenon 1990), Philippe *et al.* 1997)

One or more viruses have been isolated from 23 species of *Limacodidae* (table 3) and there can be no doubt that the list will be added to with further research. It can be seen that Picornaviruses are much more frequent in America and Africa than in Asia and that no Densoviruses have been found in the Far East. On the other hand, no *ß Nudaurelia* has been reported in Africa and America, whereas that type of virus is very frequent in Asia, along with nuclear polyhedroses and granuloses. It is not rare to find several types of virus in the same insect, and it is not always easy to determine the role of each of them in the event of an epizootic disease. These viruses have the particularity of being generally highly specific. For instance, the Densovirus type virus of ***Sibine fusca*** is also pathogenic on ***S. nesea***, but not on ***S. apicalis*** (unpublished). Epidemics are seen to occur generally in cases of overpopulation and on advanced larval instars, hence after damage. However, that is not the case with the solitary American species ***Euprosterna eleasa*** on which a nuclear polyhedrosis disease is seen to occur right from the first larval instar, and with low population levels.
It has been shown that the Hemiptera *Pentatomidae, Eocanthecona furcellata* can carry the *ß Nudaurelia* type virus pathogenic on ***Setothosea asigna*** (Sudharto, 1996). Lastly, it is possible that transmission occurs through eggs. Indeed, epidemics have been seen on colonies of ***Latoia viridissima***, transferred beforehand at the egg stage and placed in insect-proof cages (Philippe, personal communication).
It can be easy to cause epidemics artificially by spraying caterpillar populations with viral suspensions. Treatments on an industrial scale have been carried out on ***Sibine fusca*** in Colombia (Genty and Mariau, 1975a) and significant results have been obtained on other pests, such as ***Parasa lepida*** in Indonesia (Ginting and Desmier de Chenon, 1987) or ***Setothosea asigna*** (Sudharto, 1996).

Other pathogens

The literature contains numerous references to the occurrence of pathogens other than viruses, primarily fungi (table 4). With the exception of *Cordyceps* and *Beauveria,* it has to be acknowledged that these organisms do not usually occur in epidemics and, in many that these organisms do not usually occur in epidemics and, in many cases, it is reasonable to question their pathogenic nature, insofar as these organisms are able to develop *post mortem*.

Fungi of the genus *Cordyceps*, reported in Asia, mostly attack caterpillars which drop to the ground just after pupation. It is at that moment that the insects can be infected, with contamination rates sometimes exceeding 80% (Papierok *et al.*, 1993). However, when considering the development of a biological control method, the environmental conditions required for an epidemic to develop (soil type, climatic conditions, etc) are still unknown.

Table 3. Entomopathogenic viruses of *Limacodidae* on palms.

Limacodidae species	Type of virus	Origin	References
Baria elsa	?	West Africa	Mariau, 1982
Birthosea bisura	ß *Nudaurelia*	Indonesia	Desmier de Chenon *et al.*, 1989
Casphalia extranea	Densovirus	Côte d'Ivoire	Fédière, 1983; Fédière *et al.*, 1986
Darna catenatus	Nuclear polyhedrosis	Ind. (Sulawesi)	Wikardi (in Litt.), 1984
Darna nararia	Granulosis	Sri Lanka	Smith & Xeros, 1954
Darna sordida	ß *Nudaurelia* + Picornavirus	Ind. (Sumatra)	Desmier de Chenon *et al.*, 1989
Darna trima	Granulosis+ ß *Nudaurelia* Nuclear polyhedrosis Granulosis + ß *Nudaurelia* + Picornavirus	Malaysia (Sabah, Sarawak) Ind. (Sumatra) Ind. (Sumatra)	Reinganum *et al.*, 1978 Tiong & Munroe, 1977 Wikardi (in Litt.), 1984 Desmier de Chenon *et al.*, 1989
Euprosterna eleasa	Nuclear polyhedrosis	Colombia, Peru	Genty *et al.*, 1978 Mariau, 1982
Latoia pallida	Picornavirus	Côte d'Ivoire	Desmier de Chenon *et al.*, 1989
Latoia viridissima	Nuclear polyhedrosis Picornavirus (Ribovirus)	Côte d'Ivoire Côte d'Ivoire	Kouassi *et al.*, 1991 Fédière 1983; Fédière *et al., 1990* Zeddam *et al.*, 1990
Microthosea minima	Nuclear polyhedrosis	Philippines	Entwistle, 1987
Natada pucara	Nuclear polyhedrosis	Colombia	Genty *et al.*, 1978
Parasa lepida	Granulosis + ß *Nudaurelia*	Ind. (Sumatra)	Ginting and Desmier de Chenon, 1987
Setora longifurca	Picornavirus?	Phil. (Mindanao)	Entwistle, 1987

Limacodidae species	Type of virus	Origin	References
Setora nitens	*β Nudaurelia* Nuclear polyhedrosis	Mal. (Sabah) Indonesia	Greenwood and Moore, 1982 Kalshoven and Van der Laan, 1981
Setora tagalog	Nuclear polyhedrosis	Phil. (Luzon)	Cock and Godfray, 1985
Setothosea asigna	*β Nudaurelia* Cytoplasmic polyhedrosis + *β Nudaurelia*	Indonesia	Reinganum *et al.*, 1978; Sudhartho, 1996 Sinuraya & Desmier de Chenon, 1988
Sibine apicalis	Cytoplasmic polyhedrosis	Peru, Ecuador	Martignoni & Iwai, 1981
Sibine fusca	Densovirus Picornavirus	Colombia Colombia	Meynadier *et al.*, 1988; Genty & Mariau, 1973 Fédière, 1983
Sibine nesea	Densovirus and Picornavirus	Ecuador	Desmier de Chenon *et al.*, 1988
Thosea boreocaerulea	Nuclear polyhedrosis	Phil. (Luzon)	Cock & Godfray, 1985
Thosea monoloncha	Small virus? Nuclear polyhedrosis	Indonesia	Desmier de Chenon *et al.*, 1988 Lolong *et al.*, 1988
Thosea philippina	Nuclear polyhedrosis	Philippines	Entwistle, 1987

Table 4. Entomopathogenic organisms of *Limacodidae* other than viruses.

Pathogenic organism	Insects affected	References
Aspergillus flavus Link & Fries or *Aspergillus* sp.	***Parasa lepida*** ***Parasa philepida*** ***Microthosea minima*** ***Darna nararia***	Pillai & Ayyar, 1968 Evans, 1987 Holloway *et al.*, 1987 Kurian, 1963
Beauveria bassiana (Balsamo) Vuillemin	***Parasa lepida*** ***Parasa philepida***	Desmier de Chenon, 1982 Evans, 1987
Paecilomyces sp. *Paecilomyces javanicus* (Friedrichs & Bally) Brown & Smith *P. lilacinus* (Thom) Samson	***Natada subpectinata*** ***Chalcocelis albiguttatus*** ***Penthocrates rufofascia, P. styx*** ***Thosea*** sp., ***Parasaphilepida***	Genty *et al.*, 1978 Holloway *et al.*, 1987 Cock *et al.*, 1986 Evans, 1987
Verticillium lecanii	**Penthocrates styx**	
Cordyceps spp.	***Chalcocelis albiguttatus*** ***Setothosea asigna, Setora nitens, Darna bradleyi*** ***Parasa lepida, Darna mindanensis*** **(?)** ***Thosea lutea*** ***Setora cupreistriga***	Corbett, 1937 Papierok *et al.*, 1993 Wood & Nesbit, 1969 Desmier de Chenon, 1982 Schneider, 1940 Holloway *et al.*, 1987
Bacteria and fungi	***Chalcocelis albiguttatus*** ***Darna catenatus*** ***Contheyla rotunda***	Holloway *et al.*, 1987 Lever, 1969 Nirula *et al.*, 1954
Protozoa	***Setothosea asigna***	Holloway *et al.*, 1987

PYRALOIDEA

PYRALIDAE

This family includes a large number of species, which have very varied behaviours, but are very often associated with oil palm and coconut flowers.

Tirathaba rufivena Walker (Wood and Kwang Yew 1974, Cock and Godfray 1985, Corbett 1932, Paine 1935, Smee 1965, Desmier de Chenon 1985, Mariau *et al.* 1991).

Several names have been given to this pest. *T. mundella* Walker, *T. fructivora* Meyrick, *T. complexa* Butler, *Mucialla rufivena* Walker. These species have now been grouped under a single name.

Description
The adult has a wingspan of 25 mm. The wings are long and narrow, grey with a silvery appearance.
The caterpillar reaches 27 mm. It is light brown in colour, the body is shiny and covered with tactile hairs.
Pupation takes place in a sliken cocoon covered with debris.

Biology
The cycle lasts a month (incubation: 4 days; five larval instars: 16 days and pupation: 10 days).
The caterpillar develops to the detriment of coconut and oil palm flowers. On coconut, caterpillars feed first off male flowers, then perforate the female flowers and young fruits, which fall. On oil palm, the caterpillars bore into the base of fruits, which abort.
This pyralid has numerous natural enemies, the species of which vary depending on the regions.

Indonesia
Elasmus fumipennis Cameron (*Elasmidae*) on prepupae, but not very abundant. *Barichneumon muciallae* Wilk (*Ichneumonidae*) on chrysalises (10%), *Trichospilus pupivora* Ferr. (*Eulophidae*) (2%), *Telenomus tirathabae* (*Scelionidae)* Ferrière on eggs (6%), *Epitranus (Anacryptus*) *impulsator* (Wilk) (*Chalcididae*) (4%), *Apanteles tirathabae* Wilkinson on caterpillars (13%), itself parasitized by *Perilampus microgastris* Ferrière (*Perilampidae*) and *Aphanogmus manilae* Ashmead (*Ceraphronidae, Palexorista painei* Baranov (*Tachinidae*) on prepupae (0.5%), *Trichogrammatoidea nana* (Zehntner) (*Trichogrammatidae*), *Meteorus trichogrammae* Wilkinson (*Braconidae*) on caterpillars (3%), *Echthromorpha agrestoria* Swederus (*Ichneumonidae*) on chrysalises (14%), *E. tirathabae* Perk on chrysalises, *Antrocephalus renalis* Wasterston (*Chalcididae*).

The following parasitoids were successfully introduced from 1930 to 1933: *A. tirathabae*, which has itself been parasitized by *Ceraphron fijiensis* (Ferrière), *T. tirathabae,* the tachinid *Argyrophylax basifulva* Bezzi, *Devorgilla palmaris* Wilkinson and *P. painei.*

Malaysia
D. (*Venturia*) *palmaris* (*Ichneumonidae*) is the most important, *A. tirathabae*, *A. basifulva*, of little importance, was unsuccessfully introduced in Tonga, *T. pupivora.*

Philippines
A. basifulva. After its introduction in 1984, it was mass multiplied (Godfray 1985), *Apanteles tirathabae* is common and hyperparasitized by *Eurytoma* sp., *Perilampus* sp.*, A. manilae* and *Trichohalticella tirathabae* Ferrière (*Chalcididae*), *D. palmaris,* occasional, *Telenomus tirathabae* of moderate importance, *Telenomus* sp. and *Gryon* sp.*(Scelionidae).*

Papua New Guinea
Eucarcelia kokiana Towns *(Tachinidae)* on caterpillars, *Apanteles pertiades* Nixon, *Trichogramma* sp.*, Ropalidia nigra* Smith (*Vespidae*), predator on caterpillars.

Solomon islands
A. pertiades, not very important at all and *Oecophylla smaragdina* (F.) (Hymenoptera *Formicidae*), which is a predator.

Coleoneura trichogramma Meyrick (Lepesme 1947). In Vanuatu and Fiji, this species has the same behaviour as *T. rufivena.*

Hyalospila ptychis Dyar (Bondar 1940.). Similar attacks to those of ***Tirathaba*** on coconut in Brazil. Attacked fruits can nonetheless develop, but they are severely deformed.

Ephestia (Cadra) cautella Walker (Bondar 1940) on male and female coconut flowers in Brazil, and also on copra.

Hedylepta (Phostria) blackburni (Butler) (Zimmerman 1958). Parasitized by the *Ichneumonidae Trathala flavoorbitalis* (Cameron). The coconut leaf roller in Hawaii is particularly present in areas exposed to wind, maybe due to lower parasitism. Damage can be severe.

Harpagoneura complexa Butler (Lepesme 1947) attacks young coconut fruits in Brazil and Oceania.

Homoesoma sp. Coconut leaf-eater in Polynesia. Parasitized by *Apanteles* sp. and a *Chalcididae* on chrysalises and caterpillars (up to 30%); attacks are more severe in coconut plantings exposed to wind. Coconuts planted in rich soils are less severely attacked, as are those planted at high density. Defoliation can reach 40% (Julia, personal communication).

Pimelephila ghesquieri Tams (Gesquière 1931, Jover *et al.* 1951, Mariau and Morin 1971, Mariau *et al.* 1981).

Description

The adult has olive green to brown wings with a wingspan of 30 to 35 mm. The caterpillar reaches 35 mm and is dirty white to salmon pink.

Biology

The caterpillar hatches at the base of unopened leaves and mines a gallery in the leaves. On young palms, very violent attacks can be seen, and on 2 to 4-year-old palms, attacks, which are generally less violent, are clearly visible when the leaves open, and are symmetrical in relation to the rachis, forming a herring bone pattern.

Several parasitoids have been observed: *Trichospilus* sp. (*Chalcididae* on chrysalises, which reveal a polyembryonic phenomenon (50 to 100 adults per host), an *Ichneumonidae* and a *Tachinidae*.

Furcivena rhodoneuralis Hampson (Mariau *et al.* 1981). The adult measures 18 mm and the caterpillar, which reaches 20 mm, shelters beneath a very fine silken web. The cycle lasts 7 weeks.

Damage to *Elaeis* is in the form of small windows 3 to 5 cm long and 2 to 4 mm wide, which spare the upper epidermis. Quite often in Côte d'Ivoire damage is not generally serious.

The caterpillars are parasitized by an *Ichneumonidae* and the pupae by an *Eulophidae, Pediobius* sp. (frequent).

Sufetula spp.

The caterpillars of the genus attack oil palm root tips that begin to grow just above the surface of the soil. The attack stops once the root has penetrated a few centimetres into the soil, in which the caterpillar is unable to move. As attacks are localized on aerial roots, which only account for a very small proportion of the system, they do not have any impact on palm development.

S. diminutalis Walker (Genty and Mariau 1975b). In South America. There are 5 larval instars, their development taking 28 to 31 days. The females lay 50 to 80 eggs.

S. nigrescens Hampson (Mariau *et al.* 1981). In West Africa.

S. sunidesalis Walker (Desmier de Chenon 1975). The pupae are parasitized by *Antrocephalus* sp. (*Chalcididae*). A Diptera *Phoridae* can be a predator.

Two new species of ***Sufetula*** were recently discovered on coconut and oil palm roots in Sumatra in peat soils (Philippe, personal communication). In this type of soil, the caterpillars are able to move freely and attack root tips several dozen centimetres deep. In east Sumatra, a coconut planting was found where virtually all the roots of the different root categories had been attacked, leading to a substantial reduction in yields. Attacks also occur on oil palm under the same conditions (Mariau unpublished).

S. minimalis Fletcher on coconut in the Seychelles (Vesey-Fitzgerald 1941).

Elaeidiphilos (Epimorius) adustalis Hampson (Philippe 1993). In Côte d'Ivoire, the adult lays its eggs on female *Elaeis* flowers. The total cycle length is 38 to 49 days. The caterpillars destroy the stigmata, disrupting pollination. For pupation, the caterpillars weave a silken cocoon. In the event of severe attacks, the inflorescence can be entirely covered with silken threads. Attacks also occur on male flowers. Damage is very moderate.

Chevalierella elaeidis Ghesquière (Ghesquière 1943). In the Congo, the caterpillars live in *Elaeis* male flowers.

Eldana saccharina Walker (Mariau *et al.* 1981). The development cycle lasts 35 to 42 days. The female lays its eggs (450) in groups in leaf axils of young plants in oil palm and coconut nurseries. The caterpillars penetrate the plant and destroy forming tissues, which can lead to plant death. Severe attacks have been seen in Côte d'Ivoire.

Lamoria sp. attacks coconut flowers in Zanzibar (FAO 1964), ***Herculia nigrivitta*** Walker has been seen in India on dry *Cocos* flowers (Lepesme 1947), ***Antigestra*** sp., also in India (Dhileepan 1991), ***Corcyra cephalonica*** Stainton can develop on coconut fruits in Nigeria (Jerath 1965). Lastly, ***Euphys bilineata*** attacks palm fruits in Colombia (Genty, personal communication).

Mucialla rufivena Walker has been observed in India on coconut inflorescences (Thammi Raju *et al.* 1979).

BOMBYCOIDEA

SATURNIIDAE

On palms, this family is only represented by a few South American species. They are large in size and the caterpillars are covered in numerous stinging spines.

Automeris liberia Cr. (Genty *et al.* 1978).

Description

Adult: a wingspan of up to 100 mm, brownish yellow forewings and orangey hind wings with a large eye-spot thickly edged in black.
Caterpillar: cream coloured; the final instar caterpillar reaches 80 mm and is covered with long yellow spines.

Biology

A single caterpillar can destroy 4 leaflets. In Ecuador, the parasite complex includes: a *Braconidae* (*Apanteles* sp.), several Hymenoptera *Ichneumonidae Ophioninae (Enicospilus* sp.*), Pimplinae, Cryptinae, Chalcididae* on larvae, and a large Diptera *Tachinidae* on pupae.

Other species of the same genus have been observed: ***A. vividior***, ***A. bilinea*** Walker, ***A. cinctistiga*** Felder and Rogenhofer, ***Automeris*** sp. (Genty *et al.* 1978, Couturier and Kahn 1993). ***A. janus*** (Cramer), which has been reported in Trinidad and Tobago, is parasitized by the tachinid *Belvosia formosa* Aldrich.

Pseudodirphia gragatus Bouvier (Genty *et al.* 1978). Small in size: adult 75 mm and caterpillar 50 mm. The caterpillars live in colonies, primarily on young palms, pupation takes place in the soil. A *Braconidae* of the genus *Cotesia* is found on some caterpillars, along with a nuclear polyhedrosis.
Another species of the same genus has been observed in the Colombian Amazon.

Periphoba hircia (Gardiner 1967). Recently observed for the first time on oil palm in the Peruvian Amazon (Couturier and Kahn 1993). The caterpillars are gregarious and move in a procession. The female lays 200 eggs, pupation takes 40 to 50 days.

HESPERIOIDEA

All the species have been grouped in the same family.

HESPERIIDAE

This family contains numerous species, including some serious coconut and oil palm pests. The caterpillars can easily be recognized by their behaviour. Their bodies are usually smooth with a few hairs at the tip of the abdomen. They shelter in a case fomed by binding the edges of a leaflet with silken threads. This shelter is also used for pupation; the chrysalis is often covered with a white powder.

Asian species are:

Hidari irava Moore & Horsfiels (Baringbing and Bariyah 1977).

Description
The adult, with a wingspan of 42 to 54 mm, has 4 golden yellow patches on each hind wing. The caterpillar measures 46 to 53 mm.

Biology
Widespread throughout Malaysia and Indonesia, primarily on coconut; the species has also been reported on oil palm. Severe outbreaks do not occur often, but damage is nevertheless very common. The cycle lasts 50 to 56 days (5 larval instars: 32 to 37 days; pupation: 10 days). There is a substantial parasite complex. The eggs are parasitized by: *Telenomus urios* Nixon (*Neotelenomus javae* Ferrière) (*Scelionidae*), *Anastatus* sp. (*Eupelmidae)* and by *Ooencyrtus* sp. (*Encyrtidae).* Young caterpillars are attacked by *Apanteles agilis* Ashmed (*A. hidaridis* Rohw), with up to 20% parasitism and 150 individuals per host. Numerous Diptera are seen on older larvae, the most important being *Palexorista inconspicuoides* (Bar.) (*Tachinidae),* with up to 20% parasitism, but also secondarily other species such as *Erycia bezzi* Baran (*Tachinidae), Sarcophaga antilope* Boettch, *S. fuscicauda* Boettch, *S. orientaloides* S.W. *(Calliphoridae), Megaselia scalaris* Loew (*Phoridae).* Lastly, the chrysalises are attacked by: *Xanthopimpla regina* Morl. (= *X. gampsura?)* (*Ichneumonidae*), *Brachymeria euploeae* West., (up to 15%), *B. obscurata* Walker (*Chalcididae).*

Suastus gremius Fabricius (Sathiama *et al.* 1981) In India, this species can lead to 18% defoliation. The cycle lasts 36 to 47 days. The eggs are parasitized by an *Entedontidae* and an *Elachertidae* (?).

Telicota palmarum Moore (*Cephrenes chrysozona* Plotz) (Pedro *et al.* 1990). In India and Southeast Asia. The larval cycle lasts 35 days (5 instars). ***T. bambusae*** Moore in the Indo-Australian zone.

Erionata thrax L. on the Indo-Malaysian peninsula, with a development cycle of 40 to 48 days. It is parasitized by two *Elasmidae*: *Elasmus brevicornis* Gah and *E. philippinensis* Ashmead, and by an *Ichneumonidae Xanthopimpla* sp. (Corbett 1932).

Gangara thyrsis Fabricius. The caterpillar is covered by an abundant waxy secretion (Corbett 1932, Sathiamma and Abraham 1998).

Padraona chrysozona Plotz in the Philippines (FAO 1964).

American species are:

Saliana severus Mabille and ***Saliana*** sp. are very frequent on oil palm in Colombia, Peru and Ecuador, but no outbreaks have been reported. Unlike most of the species in this family, the body of the caterpillar of ***S. severus*** is covered with short hairs. Two undetermined Hymenoptera attack the eggs and a *Braconidae* the caterpillars (Genty *et al.* 1978).

Synale hylaspes Cramer on coconut in Brazil and Argentina has cycle of 40 days (Bondar 1940) and ***Perichares corydon*** Fabricius, which is more widespread (Lepesme 1947).

African and Madagascan species are:

Two species can frequently be found in oil palm plantings in West Africa (Mariau and Morin 1974, Mariau *et al.* 1981).

Pteroteinon laufella Hewitson.

Description

Adult: Shiny black body with a bluish-green sheen on the wings. The head, base of the wings and tip of the abdomen have orangey-red scales.

Caterpillar: The body is yellow to bright green. The cephalic capsule is broadly edged with black.

Biology

The species is mainly found in mature plantings. The development cycle lasts 4 to 5 weeks. A *Chalcididae, Tetrastichus* sp., parasitizes the eggs, two species of *Tachinidae* emerge from the chrysalises and two *Chalcididae* attack the caterpillars: *Brachymeria* sp. and *Stomatocerus* sp. Lastly, a highly pathogenic RNA virus affects larval populations (Herder *et al.* 1994) and a fungus of the genus *Paecilomyces* frequently affects the chrysalises.

Zophopetes cerymica Hewitson.
The adult is generally brown with a purple sheen at the tip of the wings; the hind wings have four irregular translucent patches. The cephalic capsule of the caterpillar is completely yellow ochre. This species is more common on young plantings, which can be severely defoliated.

Rhopalocampta bixae L. (Lepesme 1947). Reported on oil palm in West and Central Africa. The body and wings of the adult are blackish-brown with a purple sheen.

Pyrrhochalcia iphis Drury. (Lepesme 1947). The description of the adult is very similar to that of ***P. laufella***. Reported on coconut in western and central Africa. *Telenomus thoas* Nixon (*Scelionidae*) parasitizes the eggs.

Plötzia amygdalis Plötz. Reported on coconut in Madagascar (Morin, personal communication).

PAPILIONOIDEA

AMATHUSIIDAE

Family of Asian butterflies.

Amathusia phidippus L. (Kalshoven and van der Laan 1981, Corbett 1932, Mariau *et al.* 1991).

Description
Adult: Wingspan of 90 mm. Uniformly brown on the upperside of the wings, with two eye-spots on each of the hind wings.
Caterpillar: pale green in the young instars, then brown with a darker longitudinal line. The body has numerous long white hairs. The body terminates in two protuberances.
Chrysalis: from 40 to 50 mm, green. It is attached to the leaf by its front end.

Biology
The development cycle lasts two months (incubation: 8 days; larval cycle: 40 days; pupation: 12 days)
The natural enemies include several species of Diptera *Tachinidae*, two Hymenoptera *Chalcididae* and an *Ichneumonidae: Theronia orientalis* Gupta. (Cock personal communication).

Discophora celinde Stoll (Dammerman 1929). Reported on coconut in Indonesia (Java, Bali) with ***Pendlury*** sp.

BRASSOLIDAE

This family lives mostly off monocotyledons and is exclusively neotropical.
The species observed on oil palm and cocout all occur in Latin America. They are large and have generally brown eye-spots with a blue or purple sheen on the underside of the wings.

Brassolis sophorae L.

Description
The adult has a wingspan of over 100 mm. The caterpillar, which measures 50 mm prior to pupation, is reddish-brown with light brown longitudinal bands.

Biology (Anon. 1959, Genty *et al.* 1978)
The five to six larval instars have a development cycle lasting 50 to 85 days, and pupation lasts 10 to 14 days. The female lays 200 to 300 eggs. The caterpillars are gregarious and live grouped together during the day in a kind of silky nest made of leaflets bound together. The nests are also sometimes made in the bunches or on vegetation in the windrow. The caterpillars attack the foliage at night, and can cause spectacular defoliation. Natural enemies include a Hymenoptera *Scelionidae* of the genus *Telenomus* on the eggs, a Diptera *Tachinidae* and Hymenoptera *Chalcididae* of the genera *Spilochalcis* and *Conura* sp.*(C. morleyi* in Peru) on chrysalises; lastly, a Hymenoptera *Ichneumonidae* of the genus *Campoletis* on caterpillars, also in Peru. *Beauveria bassiana* (Bals) diseases can be seen on larvae (Habib and Andrade 1977). In the event of localized attacks, collecting and destroying the nests can help in reducing populations.

B. astyra astyra Godard (Bondar 1940). Brazilian species. It is parasitized by a tachinid *Xanthozona melanopyga* Wied (Travassos and Carrera 1941).

B. isthmia Bates (Carpenter 1943). Wide distribution from Central America to Bolivia.
No outbreaks of these last two species have ever been reported in oil palm plantings.

Opsiphanes cassina Felder (Calvache and Bautista 1998, Genty *et al.* 1978).

Description

The caterpillar is green with yellow bands. It reaches 90 mm. Like all species of the subfamily *Caligoninae*, the caterpillars have caudal appendages and cephalic horns.

Biology

The cycle lasts 59 to 77 days (eggs: 8 to 10 days; 5 larval instars: 36 to 47 days; pupation: 15 to 20 days). The eggs are laid in groups, but the caterpillars live separately. Each individual can destroy 800 cm^2 of foliage (3 leaflets).

There is a substantial parasite complex: on the eggs, *Telenomus* sp. (*Scelionidae*) and *Ooencyrtus* sp. (*Encyrtidae*); on the caterpillars: *Apanteles alius* Muesebeck and *Cotesia* sp. (*Braconidae*), *Spilochalcis nigrifrons* (Cameron) and *Brachymeria* sp. (*Chalcididae*). In addition to these parasitoids, there are predators: *Caccoleptus wicki* (Coleoptera *Dermestidae*) attacks the eggs, *Alcorynchus grandis* and *Podissus* sp. (Hemiptera *Pentatomidae*) attack the caterpillars.

The adults are strongly attracted by rotting organic matter. This characteristic is used as a control method, though it has not been clearly proved to be effective.

O. quiteria Stoll and ***O. invirae*** Hubner have numerous local forms in northern South America.

O. cassiae L. (widespread) and ***O. tamirindi*** L. have been reported in Venezuela.

Caligo nr. ***eurilochus*** Stich

Description

The adult, which has a metallic blue sheen, can reach 180 mm. The caterpillar reaches 140 mm in the final instar.

Biology (Genty *et al.* 1978)

The development cycle lasts 64 to 94 days (eggs: 7-10 days; 5 larval instars: 40 to 63 days; pupation: 17 to 21 days). A caterpillar destroys 1 500 cm^2 of foliage (5 to 6 leaflets). Outbreaks are not rare in palm plantings. The parasite complex comprises two Hymenoptera *Braconidae* (*Apanteles* sp.) and *Chalcididae* (*Brachymeria* sp.).

C. oedipus fruhstorferi Stich, with a development cycle of 70 days (Van Dinther 1956).

NYMPHALIDAE

The species belonging to this family only cause occasional damage on oil palm and coconut.

Elymnias hypermestra L. and ***E. nigrescens*** Butler in Malaysia and Indonesia, ***E. hypermestra undularis*** (Drury) and ***E. caudata*** Butler in India, ***E. fraterna*** Butler on coconut in Sri Lanka (Lever 1969, Lepesme 1947, Ganeswara Rao *et al.* 1978). Lastly ***E. nesaea*** L. in Indonesia (FAO 1964).

In Papua New Guinea ***E. agondas glaucopis*** Standinger has also been found on *Elaeis.* The cycle lasts 44 to 51 days. The species is parasitized by *Brachymeria* nr. *jambolana* Gahan and *Brachymeria* nr *lasus* Chalcididae). The caterpillars are attacked by two Hemiptera *Reduviidae, Pristhesancus femoralis* Horvarth and *Pentatomidae, Montronzeriellus melacanthus* (Boisduval) (Merrett 1993).

Hypolimnas missipus L. in the Ethiopian zone on *Elaeis* and ***Melanitis leda*** L. in Africa, Asia and Australia on *Cocos nucifera* (Lepesme 1947). Lastly, three species of the genus ***Taeneris,*** of which ***T. myops*** Felder has been observed in Papua New Guinea (Parsons 1984).

NOCTUOIDEA

NOCTUIDAE

The species of this family are often highly polyphagous. Several of them have adapted to palms, though they usually only cause limited damage.

Sesamia poephaga Tams & Bowden (Agwu 1975).
In Nigeria, the species has been observed on millet, sorghum, maize, sugar cane, etc. Damage has been reported on oil palm plantlets in nurseries, on which the insect mines a gallery in the tissues sometimes killing the plant (up to 7% mortality).

S. inferens Walker, in India.
Same damage as ***S. poephaga*** on young palms. The larval cycle lasts 55 to 62 days and pupation 12 to 14 days (Jacob and Kochu Babu 1995).

S. calamistis Hampson (Mariau *et al.* 1981).
Damage in Africa is identical to that caused by the previous species. The cycle lasts 7 to 8 weeks (eggs: 8 to 10 days; caterpillars: 4 to 5 weeks; pupation: 10 to 12 days and pre-oviposition: 2 to 3 days). Fecundity is high (up to 300 eggs per female). The eggs are laid in groups of twenty or so around the base of the petioles.

The species of the genus ***Spodoptera*** usually attack the foliage of numerous plant species.

S. litura Fabricius (Mariau *et al.* 1991). Reported in Indonesia on oil palms in the nursery. The cycle lasts 1 month with pupation taking place in the soil. Under the name ***Prodenia litura*** Fabricius (Lever 1943), this insect has been observed on coconut in New Guinea and Fiji, but the eggs would seem to have been laid accidentally.

S. exempta Walker (Yarro *et al.* 1981). On Yellow Dwarf coconut palms in Africa.

S. mauritia Boisduval (Lepesme 1947). Reported on coconut in Sri Lanka and on *Sabal* palms in Africa.

Phytometra acuta Walker (Lepesme 1947). Reported on oil palms in Asia, but also in the Congo.

Cyclodes omma Van der H. (Anonyme 1983, Mathen *et al.* 1962). The caterpillar attacks unripe coconut fruits.

Herminodes insula Dognin (Genty *et al.* 1978) South America. The wine-coloured caterpillar can reach 60 mm in length; it lives at the base of oil palm spears and browses on the tender tissues. The pupa, which is found in a fibrous cocoon at the base of young fronds, is parasitized by three species of *Ichneumonidae*. Damage is identical to that caused by ***Tiquadra*** spp. (*Tineidae*) but less serious.

Dunama angulinea. An outbreak of this species occurred in Peru on the lower fronds of 5-year-old oil palms. The substantial populations were decimated by a *Beauveria* sp. disease (Mariau unpublished). The species is common in the Colombian llanos and occurs sporadically in the Ecuadorian Amazon (Genty, personal communication).

A species of the subfamily *Nolinae* has been seen on oil palm in Côte d'Ivoire (Mariau *et al.* 1981).

LYMANTRIIDAE

This family includes serious oil palm and coconut pests. The caterpillars are covered in numerous long hairs.

Dasychira inclusa Walker (Mariau *et al.* 1991).

Serious damage has been reported in Indonesia on oil palms in the nursery and in plantations. The eggs are laid in groups of several hundred on the underside of fronds. They can be parasitized by a Hymenoptera *Scelionidae*. Young caterpillars only attack the upper epidermis, then right through the lamina. The caterpillars are parasitized by a Hymenoptera *Braconidae, Apanteles* sp. and a Diptera *Tachinidae*. The cycle lasts 51 to 57 days (incubation: 9 days; larval cycle 35 to 40 days; pupation: 8 days).

D. mendosa Hubner (Wood 1968) Highly polyphagous species. Serious damage has been reported on coconut in the state of Andhra-Pradesh, in India (Raghunath and Subramanyam 1981). Slight damage is caused on oil palm in Malaysia and Indonesia.

Calliteara horsfieldii Saunders (Mariau *et al.* 1991). The caterpillars, which can reach 5 cm in length, can cause serious defoliation on oil palm in Indonesia. Two species of Diptera *Tachinidae* emerge from the chrysalis, which is covered in white hairs. Insect activity is mostly nocturnal.

Orgyia turbata Butler (Corbett and Dover 1927, Mariau *et al.* 1991). Indo-Malaysian Peninsula. Defoliation can be severe on young palms, and sometimes on the fruits. The cycle lasts around 1 month.

Euproctis semisignata Walker (Abraham and Remamong 1976). This species bores into female coconut flowers in India. Up to 50% attacks have been observed, but the true impact on yields has not been specified. The cycle lasts 39 days (eggs: 6.5; caterpillars: 25; chrysalis 8).

E. linta Moore (Mariau *et al.* 1991) The eggs are laid in groups covered in stinging hairs. Attacks are slight on young oil palm plantings in Indonesia.

Laelia venosa Moore (Wood 1968) The caterpillar is covered in long white hairs. This is a secondary pest.

NOTODONTIDAE

Several species of this family have been observed on oil palm, sometimes with sudden serious outbreaks. The caterpillars are generally large and smooth, with highly developed anal legs.

Leptonatada sjöstedti Aurivillius (Mariau *et al.* 1981). Observed in Côte d'Ivoire. The cycle lasts 47 to 56 days (eggs: 10; 5 larval instars: 24 to 29; pupation 9 to 10). Pupation takes place on dry lower fronds. Only one outbreak has been observed, which was decimated by a viral epidemic. The rareness of outbreaks is most probably linked to a high parasitism rate (not studied) on eggs and caterpillars, and to the viral disease.

Turnaca rufisquamata Hampson. Like the previous species, episodic outbreaks have been seen in Côte d'Ivoire on oil palm. Populations can be decimated by an RNA viral disease involving the family of the *Picornaviridae* (Fédière *et al.* 1992). The same comment can be made for *Leptonatada* as regards the rareness of the species.

T. acuta Walker (Swamiappan and Sundarababu 1995, FAO 1964) attacks
coconut flowers in India.

Turnaca sp. (Wood 1968). This species, reported in Malaysia, is parasitized by a Hymenoptera *Braconidae* and a Diptera *Tachinidae*.

Arciera grisea (Holland) (Mariau *et al.* 1981). On oil palm in Côte d'Ivoire, where it is rare.

Ambadra rafflesi Moore (Mariau *et al.* 1991). Asian species reported on oil palm. At the end of its development cycle, the caterpillar measures 60 mm. The cycle lasts 1.5 months (25 days for the 5 larval instars and 10 for pupation). The eggs are parasitized by a Hymenoptera *Scelionidae*.

ARCTIIDAE

This family includes a few secondary oil palm and coconut pests in the Indo-Malaysian region. The caterpillars are covered in long hairs. (Wood 1968).

Amsacta (Creatonotus) transiens Walker, ***Asota plana*** Butler, ***Asura biseriata*** Hamps, ***Diacrisia strigatula*** Butler and ***D. strigatula rubescens*** Walker

CONCLUSION

In a well-established palm plantation, several dozen species of Lepidoptera can be found, particularly in Latin America and Asia; the entomofauna in African plantations is less rich. When combined with the associated fauna (parasitoids and predators), they form a highly complex whole, which is often developing within a fragile balance. That is why any intervention calls for the utmost care. Some species can go unnoticed for years and come to light in sudden, severe outbreaks. Such a situation may result from a decrease in a parasitoid population following human intervention or a climate change, whose consequences for insect populations oftent escape us. In some countries, considerable attention is paid to managing the populations of all species of Lepidoptera, so as to intervene as necessary at the best moment, over the minimum area required, and with highly selective insecticides.

All other things being equal, it has been seen that an abandoned plantation is attacked far less severely than a normally managed plantation (Mariau 2000). It has also often been seen that outbreaks of such leaf-eaters are much less frequent on highly scattered smallholdings than in large estate plantations, and within estates they are much rarer in border plots than in central plots. The least affected zones are in contact with a more natural environment, where biodiversity is greatest. In particular, this environment contains several plant species that secrete sweet substances in various ways, on which numerous species of parasitoid Hymenoptera feed. Based on these observations, plants of this type have recently been introduced in palm plantings. The main species belong to the families of the *Solanaceae (Solanum* spp.), *Malvaceae (Urena* sp.*), Euphorbiaceae (Croton* sp.) etc (Delvare and Genty 1992, Mariau 1993).

LEPIDOPTERA REFERENCES

Abdurahiman B.C., Mohamed U.V.K.A., Remadevi O. K (1981). Laboratory observations on the mating and oviposition behavior of *Elasmus nephantidis* Rohwer. *Philippine Journal of Coconut Studies* 6 (2): 32-33.

Abdurahiman B.C., Mohamed U.V.K.A., Remadevi O. K. (1983). Studies on the biology of *Antrocephalus hakonensis* (Hymenoptera: Chalcididae) a pupal parasitoid of *Opisina arenosella* the coconut caterpillar. *Cocos* 1: 11-16.

Abraham C.C., Remamong T.S (1976). New record of *Euproctis semisignata* Walker (*Lymantriidae – Lepidoptera*) as a serious pest of coconut inflorescence. *Current. Sci. India*: 45 (19): 707-708.

Achterberg C. van (2000). *Carinichremylus* gen. nov. (Hymenoptera: Braconidae: Pambolinae) from Peru. *Zool. Med.* 73: 251-254.

Agwu S.I.(1975) First record of *Sesamia poephagia* (*Lepidoptera, Noctuidae*) on oil palm. *Journal of NIFOR* 1975: 97-99.

Allen J.C., Bull R.A. (1984). Recent severe attack on oil palm by two caterpillars belonging to the Limacodidae. *J. W. Afri. Oil Palm Res.* 1: 130-137.

Andersson L.I.P. (1995). Aspectos generales sobre la biologia y manejo del insecto *Sagalassa valida* Walker, barrenador de las raices en la palma de aceite en Palmas de Tumaco. *Palmas* 16 (2): 17-23.

Anon. (1959). Two important pests of coconuts and their control. *The Farm Journal. British Guiana* XX (2): 10,11 et 15.

Anon. (1983). Pests and Diseses of coconut palm in Sri Lanka. Coconut Research Institute 8 p. (technical paper)

Ansari M.M., Veenakumari K., Bandyopadhyay A.K. (1992). Outbreak of *Thosea andamica* on coconut in Andaman Islands. *FAO Plant Prot. Bull.* 40 (4): 164-166.

Antony J., Kurian C. (1961). Studies on the habits and life history of *Perisierola nephantis* Muesebeck. *Ind. Coconut J.*13 (4): 145-153.

Ardila M.A., Calvache H.G. (1998). *Durrantia* sp. pos.*arcanella* (Lepidoptera: Oecophoridae), insecto inductor de la Pestalotiopsis de la palma de aceite en la zona norte. *Ceniavance Cenipalma* n° 49.

Aterrado E.D., Abad R.G. (1986). Biological studies on the white slug caterpillars, *Parasa* sp. (Lepidoptera: Limacodidae) infesting plantations in Northern Mindanao. *Philippine Journal of Coconut Studies* 11 (2): 1-8.

Austin A.D. (1987). *Ceraphronidae* in "Slug and nettle caterpillars": N.J.W. Cock, H.C.J. Godfray, J.D. Holloway eds. CAB International, Wallingford, U.K.:181-183.

Ayyar T.V.R. (1927).The parasitic Hymenoptera of economic importance noted from South India. *Bull. Ent. Res.* 18: 73-78.

Banks C.S. (1906). The principal insects attacking the coconut palm (part II). *Philippine Journal of Science* 1: 211-228.

Baringbing W.A. (1986). Biological control of the coconut moth, *Batrachedra arenosella* by *Chelonus* parasites in Indonesia. *Proceedings of the Hawaiian entomological Society* 27: 41–44.

Baringbing W.A., Bariyah B. (1977). The coconut caterpillar *Hidari irava* in Central Java, Indonesia, and screening some insecticides against it. *Journal of Plantation Crops* 5 (2): 104-108.

Bezzi M. (1925). Some Tachinidae (Dipt.) of economic importance from the Federated Malay States. *Bull. Ent. Res.* 6: 113-123.

Bodkin (1913). A new insect pest of coconut palms in British Guiana. *J. Board of Agric. British Guiana* 1 (2): 87-90.

Bondar G. (1940). Insetos nocivos e molestias do coqueiro no Brasil. *Tipografia Naval Bahia,* 160 p.

Bradley J.D. (1965). A comparative study of the coconut flat moth (*Agonoxena argaula* Meyrick) and its allies, including a new species (*Lepidoptera, Agonoxenidae*). *Bull. Ent. Res.* 56: 453-472.

Cachan P. (1959). Etude épidémiologique de la zygène (*Chalconicles catori* Jordan) des cocotiers et des palmiers à huile en Côte d'Ivoire. *J. Agric. Trop. et Bot. Appl.* 6 (2): 653-674.

Calvache H.G., Bautista P.N.F. (1998). *Opsiphanes cassina* Felder. *Ceniavances cenipalma.* 51 (Sept. 1998) 3 p.

Carpenter G. (1943). *Brassolis isthmia* Bat. *Proc. Ent. Soc. Lond.* (C) VIII: 7.

Child R. (1974). *Coconut.* London, Longmans. 335 p.

Cock M.J.W., Godfray H.C.J., (1985). Report on a programme for the biological control of selected coconut pests in the Philippines. FAO project, (ph. 1/82/003). CIBC unpublished report 64 p.

Cock M.J.W. (1987). *Chalcidoidea.* Notes on the *chalcidoids* attacking South-East Asian *Limacodidae* in "Slug and nettle caterpillars". M.J.W. Cock, H.C.J. Godfray, J.D. Holloway eds. *CAB International*, Wallingford,U.K.: 165-179.

Cock M.J.W., Gallego C.E., Godfray H.C.J. (1986). *Penthocrates* Meyrick (Lepidoptera: Limacodidae), a genus of new outbreak pests of coconut in the Philippines. Tropical Pest Management 32: 201-206.

Cock M.J.W., Godfray H.C.J., Holloway J.D. (1985). An illustrated guide to the coconut feeding *Limacodidae* of the Philippines. Ascot, U.K., Commonwealth Institute of Biological Control.

Cock M.J.W., Godfray H.C.J., Holloway J.D., eds. (1987). Slug and nettle caterpillars. CAB International, Wallingford, U.K., 270 p.

Cock M.J.W., Perera P. A.C.R. (1987). Biological control of *Opisina arenosella* Walker (Lepidoptera, Oecophoridae). *Biocontrol News and Inform.* 8 (4): 283-310.

Conway G.R., Tay E.B. (1969). Crop pests in Sabah Malaysia and their control. Sabah, Malaysia; State Ministry of Agriculture and Fisheries.

Corbett G.H. (1930). Entomological notes, first quarter 1930. Malayan Agricultural Journal 18: 212-214.

Corbett G.H. (1932). Insects of coconuts in Malaya. General Series, Department Agriculture, Straits Settlements and Federated Malay States 10, 106 pp.

Corbett G.H. (1933). Annual report 1932, Division Entomology. General Series, Department Agriculture, Straits Seetlements and Federated Malay States 14: 39-52.

Corbett G.H. (1937). Annual report 1936, Division Entomology. General Series Department Agriculture, Straits Settlements and Federated Malay States 26: 29-48.

Corbett G.H., Dover C. (1927).The life history and control of some malayan insects of economic importance I *Orgyia turbata* Butler, a possible pest of rubler *Malay. Agric. J.* 15 (7): 239-295.

Couturier G., Kahn F.(1993). A new pest of the african oil palm in the neotropics: *Periphoba hircia* (Lepidoptera Saturniidae Hemileucinae). *Principes* 37 (4): 228-229.

Crosskey R. W. (1962). A new species of *Actia* R.D. (Diptera Tachinidae) parasitic on the coconut leaf moth, *Agonoxena pyrogramma* Meyrick, in new Britain. *Bull. Ent. Res.* 53: 173-177

Dammerman K.W. (1929). The Agricultural zoology of the Malay Archipelago. Amsterdam.

Davis T.A., Sudasrip H. (1982). Causes for the defoliation of coconuts in Indonesia. *Indonesian Agriculture Research and Development Journal* 4: 8-29.

Delvare G. (1992). Les Chalcidides d'importance économique dans les palmeraies d'Amérique tropicale (Hymenoptera). *Bull. Soc. Ent. Fr.* 97(4): 349-372.

Delvare G., Genty P. (1992). Intérêt des plantes attractives pour la faune auxiliaire dans les palmeraies d'Amérique tropicale. *Oléagineux* 47 (10): 551-559.

Delvare G., La Salle J. (1993). A new genus of Tetrastichinae (Hymenoptra Eulophidae) from the Neotropical Region with the description of a new species parasitic on key pests of oil palm. *Journal of Natural History* 27: 435-444.

Desmier de Chenon R. (1975). Présence en Indonésie et Malaisie d'un lépidoptère mineur des racines du palmier à huile, *Sufetula sunidesalis* Walker et relations avec les attaques de *Ganoderma. Oléagineux* 30 (11): 448-456.

Desmier de Chenon R. (1982). *Latoia (Parasa) lepida* (Cramer) Lepidoptera Limacodidae, a coconut pest in Indonesia. *Oléagineux* 37: 177-183.

Desmier de Chenon R., (1985). Rapport Cirad.

Desmier de Chenon R., Mariau D., Monsarrat P., Fédière G., Sipayung A.. (1988). Recherches sur les agents entomopthogènes d'origine virale chez les lépidoptères défoliateurs du palmier à huile et du cocotier. *Oléagineux* 43 (3): 107-117.

Desmier de Chenon R., Sipayung A., Sudharto P.S. (1989). The importance of natural enemies on leaf-eating caterpillars in oil palm plantations in Sumatra, Indonesia. Uses and possibilities. PORIM Int. Palm Oil Conf., 5-9 Sept. 1989, Kuala Lumpur, Malaysia.

Dessart P. (1988). *Aphanogmus goniozi* sp. n., hyperparasite d'un Béthylidé au Sri Lanka (Hymenoptera-Ceraphronidea-Ceraphronidae). *Bull. Annls. Soc. r. belge* Ent. 124: 95-100.

Dharmaraju E. (1962). Parasites, the hyperparasites, predators and pathogens of the coconut leaf caterpillar *Nephantis serinopa* Meyrick, recorded in Ceylon and in India and their distribution in these countries. *Ceylon Coconut Quart.* 13 (3): 102-111.

Dhileepan K. (1991). Insects associated with oil palm in India. FAO *Plant. Path. Bull.* 39 (2-3): 94-99.

Dinther J.B.M; van (1956). Insects of the coconut palm in Surinam. *Bull. Landbouwproefstation* n° 69.

Entwistle P.F. (1987). Virus diseases of Limacodidae in "Slug and nettle caterpillars". N.J.W. Cock, H.C.J. Godfray, J.D. Holloway eds. CAB International, Wallingford, U.K.: 213-221.

Evans H.C. (1987). Fungal pathogens of Limacodidae in "Slug and nettle caterpillars". N.J.W. Cock, H.C.J. Godfray, J.D. Holloway eds. CAB International, Wallingford, U.K.: 207-224.

Fao (1964). List of insect pests of coconut palm. Second session technical working party on coconut production, protection and processing. Colombo 30/11-8/12-1964.

Fédière G. (1983). Recherches sur des viroses épizootiques de lépidoptères Limacodidae ravageurs de palmacées. Doctoral Thesis Université de Sciences et Techniques du Languedoc, France.

Fédière G. (1996). Recherches sur des viroses de lépidoptères ravageurs de cultures pérennes en Côte d'Ivoire et de cultures annuelles en Egypte. Thèse Université Montpellier II, 204 p.

Fédière G., Lery X., Kouassi N., Herder S, Veyrunes J.C., Mariau D. (1992). Etude d'un nouveau virus à ARN isolé de *Turnaca rufisquamata* (Lepidoptère *Notodontidae*) défoliateur du palmier à huile en Côte d'Ivoire. *Oléagineux* 47 (3): 107-112.

Fédière G., Monsarrat P., Mariau D., Bergoin M. (1986). A densovirus of *Casphalia extranea* (Lepidoptera: Limacodidae). Characterization and use for biological control. 4th Int. Colloquim on invertebrate pathology, August, 8-22, 1986, Veldhoven, The Netherlands.

Fédière G., Philippe R., Veyrunes J.C., Monsarrat P. (1990). Biological control of the oil palm pest *Latoia viridissima* (Lepidoptera: Limacodidae) in Côte d'Ivoire by a new Picornavirus. *Entomophaga* 35(3): 347-354.

Ganeswara Rao A. (1978). Preliminary observation on the biology of *Parena laticincta* Bates (Coleoptera: Carabidae). A predator of *Nephantis serinopa* Meyrick. *Ind. Coconut J.* 9 (1): 2-5.

Ganeswara Rao A., Thammi Raju N. B. *et al.* (1978). *Elymnias hypermestia* (Drury): a new pest of coconut in India. *Ind. Coconut J.* 9 (8): 3.

Garcia R. A., Calvache H.G., Liahernandez M., Dumar Motta N., (1994). Biologia del defoliador de la palma de aceite, *Loxotoma elegans* Zeller (Lepidoptera: Stenomidae), en Villanueva (Casanare). *Palmas* 15 (4): 9-15.

Gardiner B.O. (1967). The life history of *Periphola hircia* (Saturniidae) with a note on distriubtion and larval variation. *Journal of the Lepidopterists' Soc.* 21 (3): 198-204.

Genty P. (1972). Morphologie et biologie de *Sibine fusca* Stoll, lépidoptère défoliateur du palmier à huile en Colombie. *Oléagineux* 27 (2): 65-71.

Genty P. (1973). Observations préliminaires du lépidoptère mineur des racines du palmier à huile, *Sagalassa valida* Walker. *Oléagineux* 28 (2): 59-65.

Genty P. (1976). Etude morphologique et biologique d'un lépidoptère défoliateur du palmier à huile en Amérique latine *Darna metaleuca* Walker. *Oléagineux* 31 (2): 99-107.

Genty P. (1978). Morphologie et biologie d'un lépidoptère défoliateur du palmier à huile en Amérique latine, *Stenoma cecropia* Meyrick. *Oléagineux* 33 (8-9): 421-427.

Genty P., Desmier de Chenon. R., Morin J.P., Korytkowoski C.A. (1978). Ravageurs du palmier à huile en Amèrique Latine. *Oléagineux* 33 (7): 325-419.

Genty P., Mariau D. (1973). Les ravageurs et maladies du palmier à huile et du cocotier. Les Limacodides du genre *Sibine. Oléagineux* 28 (5): 225-227.

Genty P., Mariau D. (1975a). Morphologie et biologie du Pyralidae des racines de l'*Elaeis, Sufetula diminutalis. Oléagineux* 40 (4): 147-152.

Genty P., Mariau D. (1975b). Utilisation d'un germe entomopathogène dans la lutte contre *Sibine fusca* St. *Oléagineux* 30 (8-9): 349-354.

Ghesquière J. (1931). Sur l'importance économique et la bionomie de deux Pyraustines nouvelles pour le Congo belge, *Leucinodes orbonalis* Guen. et *Pimephila ghesquierii* Tams. *Bull. Ann. Soc. Ent. Belg.*, LXXI, p. 131.

Ghesquière J. (1943). A propos de quelques *Galleriidae* palmicoles éthiopiens et orientaux. *Rev. Zool. Bot. Afr.* XXXVII, p. 97.

Ghorpade K.D. (1976). A predacious pentatomid bug, *Cantheconidea furcellata* (Wolff) attacking *Latoia lepida* (Cramer) on mango near Bangalore. *Journal of the Bombay Natural History Society* 76: 596-599.

Ghosh S.M., Abdurahiman U.C. (1988). On the biology of *Pediobus imbreus* (Hymenoptera: Eulophidae), a hyperparasite of *Opisina arenosella*, the black-headed caterpillar of coconut. *Cocos* 6: 27-30.

Ghosh S., Chandrasekharan K., Abdurahiman U.C. (1993). The bionomics of *Pediobius imbreus* (Hymenoptera) and its impact on the biological control of the coconut caterpillar. *Journal of Ecobiology* 5 (3): 161-166.

Ginting C.U., Desmier de Chenon R. (1987). New biological prospects for controlling a major coconut pest in Indonesia: *Parasa lepida* Cramer, Limacodidae, by the use of viruses. *Oléagineux* 42: 107-118.

Godfray H. C. J. (1985). Mass rearing the tachinid fly *Argyrophylax basifulva*, a parasitoid of the greater coconut spike moth (*Tirathaba* spp.) (Lep.: Pyralidae). *Entomophaga* 3 (1): 211-215.

Gonggrijp H. (1931). Preliminary report regarding investigations on combating caterpillar pests in the oil palm cultivation. *Communication from the General Experimental Station AVROS General Series* 48, 31p.

Greathead D.J. (1987). Bombyliidae. A summary of the recorded Bombyliid parasitoids of South-East Asian Limacodidae in "Slug and nettle caterpillars": N.J.W. Cock, H.C.J. Godfray, J.D. Holloway eds., *CAB International*, Wallingford, U.K.: 195-196.

Greenwood L.K., Moore N.F. (1982). The *Nudaurelia ß* group of small RNA-containing viruses of insects: serological identification of several new isolates. *Journal of Invertebrate Pathology* 39: 407-409.

Greve J.E van S., Ismay J.W. (eds.) (1983). Crop insect survey of Papua New Guinea from July 1st 1969 to December 31st 1978. *Papua New Guinea Agricultural Journal* 32: 1-120.

Habib M.E. M., Andrade C.F. (1977). Epizootia em larvas de *Brassolis sophorae* (*Linnaeus*, 1758) causada por *Beauveria bassiana* (Bals) Vuill., com estudos de identificacao de sintomatologia. *Anais da Soc. Entom. do Brasil* 6 (2): 230-232.

Harris K.M. (1987). Tachinidae and Sarcophagidae. A summary of the recorded tachinid and sarcophagid parasitoids of South-East Asian Limacodidae in "Slug and nettle caterpillars": N.J.W. Cock, H.C.J. Godfray, J.D. Holloway eds., *CAB International*, Wallingford, U.K.: 187-193.

Herder S., Fediere G., Kouassi N., Lery X., Kouevidjin R., Philippe R., Mariau D. (1994). Mise en évidence de nouveaux virus à ARN chez *Pteroteinon laufella* Hewitson (Lepidoptera *Hesperidae*), défoliateur du palmier à huile en Côte d'Ivoire. *Oléagineux* 49 (2): 43-47.

Hering E.M. (1931). Family Limacodidae. In Seitz, A. (*ed) Die Gross-Schmetterlinge der Erde* 10: 665-728.

Hertslet L.R., Duckett J.E. (1971). *Thosea bisura* – a new major pest of oil palms. *Planter* 47, 546: 398-404.

Hinckley A.D. (1961). Insects on young coconut palms in Fiji. *Agriculture Journal* 31: 39–41.

Hinckley A.D. (1963). Parasitization of *Agonoxena argaula* Meyrick (Lepidoptera: Agonoxenidae). *Hawaiian Ent. Soc.* 18 (2): 267-273.

Holloway J.D. (1986). The moths of Borneo: key to families;families Cossidae, Metarbelidae, Ratardidae, Dudgeoneidae, Epipyropidae and Limacodidae. *Malayan Nature Journal* 40: 1-166

Holloway J.D., Cock M.J.W., Desmier de Chenon R. (1987). Systematic account of South-East Asian pest Limacodidae in "Slug and nettle caterpillars". M.J.W. Cock, H.C.J. Godfray, J.D. Holloway eds., *CAB International*, Wallingford, U.K.: 15-117.

Huguenot R., Vera J. (1981). Description et lutte contre *Castnia daedalus* Cr. (Lep. *Castniidae*), ravageur du palmier à huile en Amérique du Sud. *Oléagineux* 36 (11): 543-548.

Igbinosa I.B. (1985). Life-table studies for the nettle caterpillar. *Latoia viridissima* Holland, on the oil palm, *Elaeis guineensis* Jacq., and the coconut palm. *Cocos nucifera* L. *Agriculture, Ecosystems & Environment* 14: 77-93.

Igbinosa I.B. (1988). Parasites of the immature stages of *Latoia viridissima* Holland (Lep., Limacodidae) a pest of palms in Nigeria. *J. Appl. Ent.* 106: 527-530.

Isaac P.V. (1925). The South Andaman coconut slug-caterpillar (*Thosea unifascia,* Walk.) *Agricultural Journal of India* 20: 373-379.

Jacob S.A., Kochu Babu M. (1995). *Sesamia inferens* Walker. A new pest of oil palm seedlings in India. *Planter* 71: 265-266.

Jayanth S.P., NagarKatti S. (1984). Testing *Bessa remota* (Dip.: Tachinidae) against *Opisina arenosella*(Lep.: Crytophasidae) and other lepidopterous in India. *Entomophaga* 29 (4): 415-419.

Jerath M.L. (1965). A list of insects found on palms in Nigeria and their known parasites and predators. *Journal of NIFOR* 1965: 411-415.

Jong J.K. de (1931). *Cantheconidia furcellata* Wolff als natuurlijke vijand van rupsen in theetuinen. *Archief voor de Theecultuur in Nederlandisch-Indië* 1: 17-24.

Jover O., Galichet P., Vuillaume M.K. (1951). Sur la biologie des chenilles de *Pimelephilia ghesquieri*, parasite des palmiers à huile en pépiniere. *Rev. Path. Veget. et Entom. Agric.* 30 (3): 149-158.

Kalshoven L.G.E., van der Laan P.A. (1981). Pests of crops in Indonesia.Revised by van der Laan P.T. Ichtiar Baru, van Hoeve, Jakarta. 701 p.

Kalshoven L.G.E., van der Vecht J. (1950). De plagen van de cultuurgewassen in Indonesie. Deel I. Gravenhage/Badoeng; N.V. Uitgeverij W. van Hoeve.

Kapadia M.N. (1981). Susceptibility of three cultivated coconut varieties to *Nephantis serinopa* Meyrick. *Indian Coconut Journal* 12: 56.

Kapadia M.N. (1983). Record of *Pediobius* spp. as hyperparasites of *Perisierola nephantidis* M. and *Bracon brevicornis* W. in Gujarat. *Indian Coconut Journal* 14: 7.

Kerrich G.J. (1961). A new australasian species of *Elachertus spinola* (Hym., Chalcidoidea, Eulophidae) and notes on two european species. *Bull. Ent. Res.* 52: 769-772.

Kimura N. (1978). A new nettle caterpillar of oil palm in Sabah, Malaysia. *Japan Agricultural Research Quarterly* 12: 53-55.

Korytkowski C.A., Ruiz A.E.R.(1980). El barreno de los racimos de la palma aceitera, *Castnia daedalus* (Cramer), en la plantacion de Tocache (Peru). *Oléagineux* 35 (1): 1-11.

Kouassi N.K., Fédière G., Lery X., Philippe R., Bergoin M. (1991). Detection of a nuclear polyhedrosis Baculoviruses in *Latoia viridissima* Limacodid Lepidoptera, an oil palm and coconut pest in Côte d'Ivoire. *Oléagineux* 46(2): 53-59.

Krishnan R. (1977). Larval biology of *Crematopsyche pendula. Planter* 53 (618): 381-394.

Kurian C. (1954). Descriptions of some new chalcids (parasitic Hymenoptera) from India. *Agra University Journal of Research* 3: 119-128.

Kurian C., Sathiamma B., (1970). Leaf-eating caterpillar a problem in coconut cultivation. *Indian Farming* 20 (8): 17-18.

Lepesme P. (1947). Les insectes des palmiers. Paul Lechevalier Ed. Paris. 903 p.

Lever R.J.A.W. (1937). Economic Insects and Biological Control in the British Solomon Islands. *Bull. Ent. Res.* 28: 325-331.

Lever R.J.A.W. (1943). Entolomogical Services in Fidji. *Bull. Ent. Res.* XIV: 92.

Lever R.J.A.W. (1953) Notes on outbreaks, parasites, and habits of the coconut moth (*Artona catoxantha*). *Malayan Agric. Journal* 36 (1): 20-27.

Lever R.J.A.W. (1964) Notes sur quelques parasites, hyperparasites et prédateurs des insectes s'attaquant au cocotier en Malaisie. *Bull. phytosan. FAO* 12 (2): 43-44.

Lever R.J.A.W. (1969). Les ravageurs du cocotier. Etudes agricoles de la FAO n° 77. FAO Rome, 196 p.

Lolong A., Zelazny B., Mawikere J. (1988). Studies on *Thosea monoloncha* integrated coconut pest control. Project Annual Report. Coconut Res. Inst. Manado, North Sulawesi Ind.: 79-83.

Mac Quillan M.J. (1969). Observations on a borer of coconut petioles in the british Solomon island protecterat. *PNG Agric. J.* 20 (3-4): 71-74.

Manolin Avila I.A. (1986). Morfologia y biologia de un defoliador de la palma africana: *Oiketicus kirbyi.* FAO mesa redonda sobre Palma aceitera. Junio 1986 Valledupar (Colombia).

Mariau D. (1982). Les ravageurs phyllophages du palmier à huile et du cocotier. Importance des parasites entomopathogènes dans la régulation des populations. *Oléagineux* 37(1): 3-7.

Mariau D. (1993). Lutte intégrée dans les palmeraies: bilan. *Oléagineux* 48 (7): 309-318.

Mariau D. (1999). Les *Limacodidae (Lepidoptera)* du palmier à huile et du cocotier. Espèces nuisibles et ennemis naturels. *Plantations, recherche, développement* 6 (3): 149-160.

Mariau D. (2000). Problèmes entomologiques en replantation des palmeraies et des cocoteraies. *OCL* 7 (2): 203-206.

Mariau D., Desmier de Chenon R. (1990). Importance du rôle des virus entomopathogènes dans les populations de lépidoptères défoliateurs des palmiers. *Oléagineux* 45 (11): 487– 491.

Mariau D., Desmier de Chenon R., Julia J.F. (1981). Les ravageurs du palmier à huile et du cocotier en Afrique occidentale. *Oléagineux* 36 (4): 169-228.

Mariau D., Desmier de Chenon R., Sudharto P.S. (1991). The insect pests of oil palm and their enemies in Southeast Asia. *Oléagineux* 46 (11): 400-476.

Mariau D., Huguenot R. (1983). Méthode d'estimation des populations de *Castnia daedalus* (Lépidoptère *Castniidae*) sur le palmier à huile. *Oléagineux* 36 (11): 543-548.

Mariau D., Julia F.J. (1973). Les ravageurs et maladies du palmier à huile et du cocotier. Les *Parasa. Oléagineux* 28(3): 129-131.

Mariau D., Morin J.P. (1971) Les maladies et ravageurs du palmier et du cocotier. La pyrale du palmier à huile. *Oléagineux* 26 (6): 379-381.

Mariau D., Morin J.P. (1974). Les ravageurs et maladies du palmier et du cocotier. Chenilles de la famille des Hesperidae en Afrique occidentale. *Oléagineux* 29 (4): 187-188.

Martignoni M.E., Iwai P.J. (1981). A catalogue of viral diseases of insects, mites and ticks. In Burgess, E.D. (ed.) Microbial control of pests and diseases. London and New York, Academic Press: 897-911.

Mathen K., Pillai G.B., Antony J., Kurian C. (1962). Less serious enemies of the coconut palm. *Coconut Bulletin* 16 (4): 99-100.

Mathen K., Sathiamma B., Chandy Kurian (1968). A mite to combat coconut caterpillar. *Indian Farming* August 1968: 33-34.

Mathewkutty T.I., Subramanian K.V. (1997). Outbreak of *Contheyla rotunda* H. on coconut in Periyar district of Tamilanadu. *Indian Coconut Journal* 28 (2): 2-3.

Menon K.P.V., Pandalai K.M. (1958). The coconut palm. A monograph. *Indian Central Coconut Committee* Ernakulam S. India 384 p.

Merrett P. J. (1993). Life history of *Elymnias agondas glaucopis* (Nymphalidae: Satyrinae) a pest of oil palm in Papua New Guinea. *J. of the Lepidopterists' Society* 47 (3): 229-235.

Meynadier G., Amargier A., Genty P. (1977). Une virose de type densonucléose chez le lépidoptère *Sibine fusca* Stoll. *Oléagineux* 32 (8-9): 357-361.

Miller N.C.E. (1929). Notes on *Setora nitens* Walk., a nettle caterpillar pest of coconut palms, with special reference to outbreaks in the Teluk Anson and Bagon Datoh Districts. *Malayan Agricultural Journal* 17: 315-25.

Minet J. (1989) Remaniement partiel de la classification des Gelechioidea, essentiellement en fonction de caractères pré-imaginaux (Lepidoptera Ditrysia). *Alexanor* 16 (4): 239-255.

Mohandas T.P., Abdurahiman U.C. (1994). Effect of temperature on the reproductive biology of *Antrocephalus hakonensis* (Ashmead) (Hymenoptera: Chalcididae), a pupal parasitoid of *Opisina arenosella* Walker, the caterpillar pest of coconut. *J. of Entomological Research* 18 (3): 203-211.

Murthy K.S.,, Gour T.B., Reddy D.D.R., Babu T.R., Zaheruddeen S.M. (1995). Pheromone baits for coconut black heade caterpillar, *Opisina arenosella* Walker. *Insect Environment* 1: 3-4.

Nadarajan L., Channabasavanna G. P. (1978). Effect of temperature and humidity on the biology and development of the parasites *Perisierola nephantidis* Muesebeck and *Tetratichus israeli* Mani and Kurian. *Bull. Ent. Res.* 19: 188-196.

Nadarajan L., Channa Basavanna G.P. (1980a). Population dynamics of coconut black headed caterpillar, *Nephantis serinopa* Meyrick (Lepidoptera: Cryptophasidae) and its parasites. *Mysore J. agric. Sci.* 14: 533-541.

Nadarajan L., Channa Basavanna G.P. (1980b). Studies on the integrated control of coconut blackheaded caterpillar *Nephantis serinopa* Meyrick (Lepidoptera: Cryptophasidae). *University of Agriculture Sciences Tech. Series* No. 43, 34 p.

Nadarajan L., Channa Basavanna G.P. (1980c). Sterilisation of the coconut black headed caterpilar *Nephantis serinopa* Meyrick. *Indian J. Ent.* 42 (4): 680-689.

Nadarajan L., Jayaraj S. (1975). Influence of various hosts on the development and reproduction of the pupal parasite, *Tetrastichus israeli* M. and K. (Eulophidae: Hymenoptera). *Curr. Sci.* 44 (13): 458-460.

Narendran J.C. (1985). A taxonomic revision of the chalcid parasites (Hymenoptera/Chalcididae) associated with *Opisina arenosella* Walker (Lepidoptera *Xylorictidae*). *Entomon.* 10 (2): 83-96.

Nirula K. K.(1955). Important pests of the Coconut palm: Check list. *Ind. Coconut J.* 8 (3): 1-13.

Nirula K.K., Antony J., Sahasranaman K.N., Menon K.P.V. (1954). Observations on the biology and control of *Contheyla rotunda* H. *Ind.Coconut J.* 7: 143-155.

Ooi A.C.P. (1978). The integrated control of the coconut leaf moth, *Artona catoxantha* Hamps, 8 p.

Ortiz S.L.E., Calvache H., Luque E.Z. (1994) Control microbiano de *Sagalassa valida* Walter (Lepidoptera: Glyphipterigidae) con el nematodo *Steinernema carpocapsae* (Weiser) en Tumaco. *Palmas* 15 (1): 29-37.

Paine R.W. (1935). The control of the coconut spike moth, (*Tirathaba trichogramma* Meyr) in Fiji. *Fidji Depart. of Agric. Bull.* 18: 1-30.

Papierok B., Desmier de Chenon R., Freulard J.M., Suwandi W.P. (1993). New perspectives in the use of a *Cordyceps* fungus in biological control of nettle caterpillars in oil palm plantations. 1993 PORIM Int. Palm oil Congress, 20-25 Sept., Kuala Lumpur, Malaysia, 6 p.

Parsons M. (1984). Life histories of *Taenaris* (*Nymphalidae*) from Papua New Guinea. *Journal of the Lepidopterists' Society* 38 (2): 69-84.

Pedro L.B., Borines L.M. de, Gapasin D.P. (1990). Biology and host range of the coconut leaffolder, *Telicota palmarum* Moore (Hespiriidae, Lepidoptera). *Annals of Tropical Research* 1990: 10-17.

Perera P.A.C.R., Hassell M.P., Godfray H.C.J. (1988). Population dynamics of the coconut caterpillar, *Opisina arenosella* Walker (Lepidoptera: Xylorictidae), in Sri Lanka. *Bull. Ent. Res.* 78: 479-492.

Peter C. (1987). Parasites of *Bessa remota* Aldrich, *FAO Plant. Prot. Bull.* 35 (2): 63.

Philip B.M., Mathai S., Jacob A. (1982). A nuclear polyhedrosis of the black headed caterpillar, *Nephantis serinopa* Meyrick (Lepidoptera: Crytophasidae). *Current Science* 51 (12): 611.

Philippe R. (1993). Etude de l'incidence des ravageurs sur les inflorescences femelles du palmier à huile en Afrique de l'Ouest. *Oléagineux* 48 (10): 389-405.

Philippe R., Veyrunes J.C., Mariau D., Bergoin M. (1997). Lutte biologique avec des virus entomopathogènes. Application aux ravageurs du palmier à huile et du cocotier. *Plantations, recherche, développement* 4 (1): 39-45.

Piepers M.C., Snellen P.C.T. (1900). Enumération des lépidoptères Heterocéres recueillis à Java. *Tijdschrift voor entomologi* 43: 12-106.

Pillai K.S., Ayyar G.R. (1968). Biology of *Parasa lepida* Cram., a pest of coconut in Kerala. *Agricultural Research Journal Kerala* 6: 101-104.

Pillai G.B., Kesava Bhat S. (1987). Biology and feeding potential of *Parena nigrolineata* (Chaudoir) (Coleoptera: Carabidae), a predator of the coconut caterpillar *Opisina arenosella* Walker. *Ind. Coconut J.* 17 (2): 6-10.

Pillai G.B., Nair K.R.(1987a). Observations on *Xanthopimpla sp.* (Hymenoptera: Ichneumonidae), a pupal parasite of *Opisina arenosella* Walker, on coconut in Kerala. *J. of Biological Control* 1 (1): 7-10.

Pillai G.B., Nair K.R. (1987b) *Brachycoryphus nursei* (Cameron) (Hymenoptera: Ichneumonidae), a parasitoid of *Opisina arenosella* Wlk. on coconut in Kerala, India. *Journal of Plantation Crops* 15 (1): 9-14.

Pillai G.B., Nair R.K. (1993). Studies on the chalcidid pupal parasitoids of the coconut caterpillar *Opisina arenosella* Walker in Kerala, India. *Entomon.* 17 (3-4): 183-192.

Ponto S.A.S., Mo T.T. (1950). De bestrijding van een slakrups, *Darna (Orthocraspeda) catenatus* (Sn.), op klapper in het Palu-dal. *Landbouw Buitenzorg* 22: 69-81.

Polaszek A.B. (1987). Chrysididae. A summary of the records of Chrysidid parasitoids of South-East Asian Limacodidae in "Slug and nettle caterpillars". N.J.W. Cock., H.C.J. Godfray, J.D. Holloway eds., *CAB International*, Wallingford U.K.: 185-186.

Pushpalatha N.A., Veeresh G.K. (1995). Numerical relationship beween *Opisina arenosella* Walker and its natural enemies, *Apanteles taragmae* Vierick and *Parena nigrolineata* Chaudoir on coconut. *J. insect Sci.* 8 (2): 148-150.

Raghunath T.A.V.S., Subramanyam B.H (1981). On the occurence of the hairy caterpillar *Dasychira mendosa* (Hybner) on coconut. *Ind. Coconut J.* 12 (4): 4.

Raghwani B.R., Kapadia M.N., Butani P.G. (1997). Effectiveness of the pupal parasitoids, *Brachymeria nephantidis* Gahan and *Brachymeria lasus* Walker against *Opisina arenosella* Walker. *Journal of Oil seeds Research* 14 (1): 116-117.

Ramachandran C.P., Ponnamma K. N., Koya K.M.A., Kurian C. (1979). The coconut leaf-eating caterpillar, *Nephantis serinopa* Meyrick, a review. *The Philipp. J. Coconut Studies* 4 (2): 9-17.

Rao V.P. (1964). Natural enemies of the insect pests of coconut. *FAO Cong. Colombo* 12-1964.

Rao V.P, Sudha Rao V. (1964). *Spoggosia (Stomatomyia) bezziana* (Bar.) a Tachinid parasite of the coconut caterpillar *Nephantis serinopa* Meyrick. *Tech. Bull. Commonwealth Inst. Biol. Control* 4: 95-112.

Rapp G. (1989). Coconut Stick insect *Graeffea crouanii (Orthoptera: Phasmidae*) Coconut flat moth *Agonoxena argaula* (*Lepidoptera: Agonoxenidae*). Studies on pest status, biology, ecology and control carried out in the Kingdom of Tonga, South Pacific. Stuttgart-Hohenheim, 113 p.

Reinganum C., Robertson J.S., Tinsley T.W. (1978). A new group of RNA viruses from insects. *Journal of General Virology* 40: 195-202.

Rhainds M., Gries G.,Rodriguez R. (1995). Evidence for mate choice by male bagworms, *Oiketicus kirbyi* (Guilding) (Lepidoptera: Psychidae). *The Canadian Entomologist* 127 (6): 799-803.

Rhainds M., Gries G., Saleh A. (1997). Sex specific habitat utilization by bagworms (Lepidoptera: Psychidae). *The Canadian entomologist* 129 (1): 119-200.

Rhainds M., Gries G.,Saleh A. (1998). Density and pupation site of apterous female bagworms, *Metisa plana* (Lepidoptera: Psychidae), influence the distribution of emergent larvae. *The Canadian Entomologist* 130: 603-613.

Roepke W. (1935). De Slakrupsenplaag op het Molukken-eiland Batjan. Mededelingen van de Landbouwhoogeschool te Wageningen 39: 40 p.

Ruiz E.R.A., Korytkowski C.A. (1979). *Oencyrtus* sp. (*Hymenoptera Encyrtidae)* parasito de huevos de *Castnia dedalus* (Cramer) *(Lepidoptera Castniidae) Revista Peruana de Entomologia* 22 (11): 13-15.

Sankaran T. (1970). The oil palm bagworms of Sabah and the possibilities of their biological control. *PANS* 16 (1): 43-55.

Sankaran T., Syed S.R.A. (1972). The natural enemies of bagworms on oil palms in Sabah, East Malaysia. *Pacific Insects* 14 (1): 57-71.

Sasaerila Y., Gries G., Khaskin G., Gries R., Hardi (1997). Identification of sex pheromone components of nettle caterpillar, *Sethothosea asigna. Journal of Chemical Ecology* 23 (9): 2187-2196.

Sasaerila Y., Gries R., Gries G., Khaskin G., King Skip, Boo Teo Chor. (2000a). Decadienoates: Sex pheromone components of nettle caterpillars *Darna trima* and *D. bradleyi. Journal of Chemical Ecology* 26 (8): 1969-1981.

Sasaerila Y., Gries G., Gries R., Khaskin G., Hardi (2000b). Sex pheromone components of nettle caterpillar, *Setora nitens. Journal of Chemical Ecology* 26 (8): 1983-1990.

Sathiamma B., Abraham V.A. (1998). Observations on the leaf roller, *Gangara thyrsis* Moore a pest of coconut palm in India. *Indian Coconut Journal* 29 (1): 8-11.

Sathiamma B., Abraham V.A., Chandikurian (1981). The leaf roller *Suastus gremius* Fab. (Lepidoptera: Hesperiidae), a minor pest of coconut palm. *Ind. Coconut* J. 12 (2): 6-7.

Sathiamma B., Jayapal S.P. Pillai G.B. (1985). Record of *Ankylopteryx octopunctata candida* (Fabricius) (Neuroptera: Chrysopidae), as egg and larval predator on *Opisina arenosella* Wlk., the leaf eating caterpillar of the coconut palm. *Current science* 54 (21): 1128.

Sathiamma B., Jayapal S.P., Pillai G.B. (1987). Observations on spiders (order: Araneae) predacious on the coconut leaf eating caterpillar *Opisina arenosella* Wlk. (*Nephantis serinopa* Meyrick) in Kerala: biology of *Rhene indicus* tikader (Salticidae) and *Cheiracanthium sp.* (Clubionidae). *Entomon.* 12 (2): 121-126.

Sathiamma B., Pillai G.B., Jayapal S.P. (1986). Bioecology of *Bracon hebetor* Say (Hymenoptera: Braconidae), an indigenous larval parasitoid of the coconut laf eating caterpillar *Opisina arenosella* Wlk. (Lepidoptera: Cryptophasidae). *Journal of Plantation Crops* 14 (2): 140-146.

Schneider F. (1940). Schädinskten und ihre Bekämpfung in Ostindischen Gambirkulturen. *Mitteilungen der Schwiezerischen Entomologischen Gessellschaft* 18: 77-207.

Schuilling M., Van Dinther J.B.M.. (1980). Ecology and control of *Castnia dedalus*, a major pest of oil palm in Brazil. *Zeitschrift fur Angewandte Entomologi* 90 (2): 161-174.

Segeren H. A. (1984). Biological control of *Castnia dedalus* with insect parasitic nematodes. *Surinaamse Landbouw.* 32 (2): 45-50.

Shivashankar T, Annadurai R.S., Srivinas M., Preethi G., Sharada T.B., Paramashivappa R., Srinivasa Rao A., Prabhu K.S., Ramadoss C.S., Veeresh G.K., Subba Rao P.V. (2000). Control of coconut black-headed caterpillar (*Opisina arenosella* Walker) by systemic application of "Soluneem" – A new water-soluble neem insecticide formulation. *Current Science* 78 (2): 176-179.

Simmonds H.W. (1924). Report on mission to New Guinea, Bismarcks, Solomons and New Hebrides. Suva. Legislative Council Paper, Fiji.

Simmonds H.W. (1938). Coconut pests and diseases, *Fiji Department of Agriculture Bulletin* n° 20.

Simmonds H. W. (1951). Premature nutfall on Taveuni. *Agricultural Journal* 22 (1): 22-24.

Singh SH. (1953). Notes on the life history on the coconut leaf moth *Agonoxena argaula* Meyr. *Fiji Agric. J.* 24: 3-4.

Sinuraya L., Desmier de Chenon R. (1988). Biological control of an oil palm pest in North Sumatra *Setothosea asigna* Van Eecke,Lep. *Limacodidae* with the use of a viral solution. Symp. on biol. Control of pests in tropical agriculture ecosystems. Bogor, Indonesia, June 1-3, 1988: 85-94.

Sipayung A., Desmier de Chenon R., Sudharto P. (1989). Natural enemies of leaf-eating Lepidoptera in oil palm plantations, North Sumatra. *Biotrop Spec. Publ.* 36: 99-121.

Smee L. (1965). Insect pests of *Cocos nucifera* in the territory of Papua New Guinea:their habits and control. *Papua New Guinea Agric. J.* 17 (2): 53-64.

Smith K.M., Xeros (1954). A comparative study of different types of viruses and their capsules in the polyhedroses and granuloses of insects. *Parasitology* 44: 400-406.

Soekarjoto S., Sudasrip H., Davis T.A. (1980). *Setora nitens*, a serious sporadic insect pest of coconut in Indonesia. *Planter.* 56: 167-182.

Sudharto Ps. (1996). Etude d'un virus pathogène de *Setothosea asigna* van Eecke (Lepidoptera:Limacodidae), ravageur du palmier à huile en Indonésie: caractérisation, diagnostic et épidémiologie en plantation. Thèse de Doctorat, Université Montpellier II, 146 p.

Sudharto Ps., Desmier de Chenon R., Sipayung A. (1990). Leaf eating caterpillars of secondary importance on oil palm, a potential risk in plantations. 32d Int. Conf. on Plant Protection in the tropics. Malaysian Plant Prot. Society, 20-23 March 1990, Genting Hifhlands Pahang, Malaysia.

Sujatha A, Singh S.P. (1999). Natural enemy complex of coconut leaf eating caterpillar, *Opisina arenosella* Walker (Lepidoptera: Xylorictidae) in Karnataka. *Journal of Biological Control.* 13 (1-2): 51-58.

Swamiappan M., Sundarababu P.C. (1995). Outbreak of a new leaf caterpillar, *Turnaca acuta* Wlk. (*Notodontidae*: Lepidoptera) on coconut in Tamil Nadu. *Ind. Coconut J.* 25 (1): 2-3.

Syed R.A. (1978). Bionomics of the three important species of bagworms on oilpalm. *Malaysian Journal of Agriculture.* 51 (4): 392-398.

Syed R.A., Pang T.C. (1972). Status, history and control of *Setora nitens.* On Cocoa and coconuts in Malaysia. Kuala Lumpur, Incorporated Society of Planters: 322-328.

Syed R. A., Saleh H.A. (1998). Integrated pest management of bagworms in oil palm plantations of PT.PP London Sumatra Indonesia TBK (with particular reference to *Mahasena corbetti* Tams) in North Sumatra. *1998 International Oil Palm Conference* September 23-25 1998, 7 p.

Syed R.A., Shah Salman, Liew k.T. (1974). Occurrence and control of *Mahasena corbetti* Tams in oil palms in Sabah. *Advances in oil palm cultivation* Ed. R.L. Wastic and D.A.Earp. 402-414. Incorporated Society of Planters. Kuala Lumpur.

Tams W.H.T. (1931). Four moths of the family Limacodidae injurious to coconut palms. *Bull. Ent. Res.* 21: 489-490.

Taylor T.H.C. (1930). Early nutfall from coconut palms in Fiji. *Fiji Department of Agriculture Bulletin* n° 17.

Thammi Raju N.B., Ganeswara Rao A.., Lakshmi Narayana K. (1979). *Mucialla rufivena* Walker (Pyralidae Lepidoptera) a new inflorescence pest of coconut. *Indian Coconut J.* 10 (2): 4-5.

Tiong R.H.C. (1979). Some predators and parasites of *Mahasena corbetti* (Tams) and *Thosea asigna* (Moore) in Sarawak. *Planter* 55: 279-289.

Tiong R.H.C. (1982). Study of some aspects of biology and control of *Thosea asigna* (Moore). In Pushparajah, E., Chew Poh Soon (eds). The oil palm inthe eighties. Kuala Lumpur, Incorporated Society of Planters, pp. 529-542.

Tiong R.H.C., Munroe D.D. (1977). Microbial control of an outbreak of *Darna trima* (Moore) on oil palm (*Elaeis guineensis* Jacq.) in Sarawak (Malaysian Borneo). In Earp, D.A., Newall W. (eds). International developments in oil palm. Kuala Lumpur, Malaysia, Incorporated Society of Planters: 624-639.

Tothill J.D., Taylor T.H.C., Paine R.W. (1930). The coconut moth in Fiji. A history of its control by means of parasites. London, Imperial institute of Entomology.1930). (Diptera Tachinidae), praga das palmeiras. Dados bionomicos dos dois insetos et morfologicos do taquinideo. *Arc. Zool. S. Paulo* III: 43.

Travassos L., Carrera M. (1941). *Xanthozona melanopyga* (Wiedmann), 1930). (Diptera Tachinidae), praga das palmeiras. Dados bionomicos dos dois insetos et morfologicos do taquinideo. *Arc. Zool. S. Paulo* III: 43.

Vecht J van der. (1950). The coconut leaf moth (*Artona catoxantha* Hamps) Pt. I. Life history and habits of its parasites and hyperparasites.*Colon. Plant and Animal Prod.* 2 (2): 140.

Vesey-Fitzgerald D. (1941). Some insects of economic importance in Seychelles. *Bull. Ent. Research* 32: 158.

Wahid M. B., Halim Hassan H.J. A., Masijan Z. (1988). Bagworms (Lepidoptera: Psychidae) of oil palms in Malaysia. *PORIM occasionnal paper* n° 23.

Wahid M.B., Kamaruddin N. (1993). The population dynamics of the bag worm *Metisa plana* (Lepidopterae: Psychidae) and it associated natural enemies on oil palm at Porim Kluang Malaysia. *Porim Int. Palm oil Congress* 2-25 Sept. 1993. Kuala Lumpur. Malaysia.

Walker A.K. (1989). Synonymy of two *Apanteles* species (Hymenoptera: Braconidae) parasitizing the flat moth, *Agonoxena argaula* Meyrick (Lepidoptera: Agonoxenidae). *Bull. Ent. Res.* 79: 283-284.

Way M.J., Cammel M.E. (1989). Ants (Hymenoptera: Formicidae) as egg predators of coconut pests, especially in relation to biological control of the coconut caterpillar, *Opisina arenosella* Walker (Lepidoptera: Xyloryctidae), in Sri Lanka. *Bull. Ent. Res.* 79 (2): 219-233

Wood (1968). Pests of oil palms in Malaysia and their control. *The incorporated society of planters,* Kuala Lumpur, Malaysia: 204 p.

Wood B.J., Kwang Yew (1974). Studies on the biology and control of the oil palm bunch moth, *Tirathaba mundella* (Walker) *Lepidoptera: Pyralidae. Malaysian Agric. J.* 49 (3): 310-311.

Wood B.J., Nesbit D.P. (1969). Caterpillar outbreaks on oil palms in Eastern Sabah. Planter, Kuala Lumpur 45: 285-299.

Wolcott G.N. (1933). An economic Entomology of the West Indies. *Porto Rico.*

Yadav D.N. (1985). Influence of nectar on the fecundity, longevity and sex-ratio of *Bracon brevicornis* Wems, an important parasite of *Opisina arenosella* Walk., the blackheaded caterpillar of coconut. *Current Science,* India 54 (17): 877-878.

Yarro J.G., Otindo B.L., Gatehouse A.G, Lubega M.C. (1981). Dwarf variety of coconut, *Cocos nucifera* (Palmae), a hostplant for the African armyworm, *Spodoptera exempta* (Walk.) (*Lepidoptera: Noctuidae*). *Insect Science and its Applic.* 1 (4): 361-362.

Zagatti, P., Desmier de Chenon R., Genty Ph., Rochat D., Mariau D. (1993). The pheromonal substances of some oil palm insect pest species: initial results. 1993 Pipoc Palmoil Congress "update and vision" 20-25/9 Kuala Lumpur Malaysia: 38.

Zeddam J.L., Philippe R., Veyrunes J.C., Fédière G., Mariau D., Bergoin M. (1990). Study of the Ribovirus of *Latoia viridissima* Holland, a palm pest in West Africa. Characterization. Diagnosis. Serology and epidemiological monitoring. *Oléagineux* 45 (11): 493-500.

Zimmerman E.C. (1958). Insects of Hawai. 8, Lepidoptera: Pyraloidea. Honolulu, University of Hawai.

HEMIPTERA

INTRODUCTION

In 1947, Lepesme inventoried 167 species of Hemiptera living off the coconut palm *Cocos nucifera* and 74 off the oil palm *Elaeis guineensis*, though a large number of them were common to both palms. In half a century of research, that number has increased substantially and the current figure is probably twice or three times as large (Mariau, 1998).

The species of Hemiptera found on palms can be divided into three major groups:

Fulgoromorpha and *Cicadomorpha*: which include a very large number of species that puncture the leaves. Many of these species only feed off palms in their adult state and probably, for many of them, in an episodic way. None of these species has proved to be a serious primary pest, but several transmit diseases.

Sternorhyncha, which includes numerous species of scale insects, aphids and whiteflies. Relatively few of the insects are serious pests, but some of them can occur in spectacular outbreaks.

Heteroptera, in which several species of bugs are major depredators and others are disease vectors; lastly, a few species are predators on insects that attack palms.

HEMIPTERA-FULGOROMORPHA

Studies undertaken on the transmission of oil palm and coconut diseases in Africa, Asia and Latin America have revealed a very large number of species more or less living off palms that are suspected of acting as vectors for various diseases. For instance, Louise and Filho (1993) inventoried more than 500 species on oil palm in Brazil, the most represented species being the *Cicadellidae* (50%), *Derbidae* (19%) and *Membracidae* (13%). A similar study in Ecuador by Perthuis (1994) led to an inventory of more than 250 species on oil palm leaves, the most represented families being the same as in Brazil. Using labelled markers injected into the plant, Eskafi (1982) mentions 28 species of planthoppers and leafhoppers on coconut in Jamaica.

However, a small number of species is able to multiply on coconut. Several of them have proved to be directly involved in transmitting diseases of varied etiologies.

DELPHACIDAE

Two species: ***Sogatella kolophon*** Kikaldy and ***Tagasodes cubanus*** Crowford, are vectors of the viral-like disease known as bud rot in Côte d'Ivoire (Julia and Mariau 1982). Also in Africa, the species ***Nesodryas antiope*** Fennah was seen on coconut in Mauritius (Fennah 1964), then in Tanzania (Wilson 1987). Feeding sites of ***Peregrinus maidis*** (Ashmead), a species living off maize, have been described on coconut in Florida (Tsai and Fisher 1993). In Brazil, 18 undetermined species have been observed on *Elaeis* (Louise and Filho1993) and 6 in Ecuador (Perthuis 1994).

DERBIDAE

A very large number of species feeds off oil palm and coconut. In Sierra Leone, Hargreaves (1937) inventoried 23 species mostly belonging to the genera ***Diostrombus, Lydda*** and ***Zoraïda***. In Ghana, numerous species belonging to the genera ***Kamendaka, Lydda, Diostrombus***, and in Nigeria to the genera ***Zorabana, Zoraïda, Pamendanga***, etc, have been observed on coconut (Dery *et al.*, 1996). Wilson (1987b) also inventoried 35 species on palms in Africa, including 7 new species on coconut in Tanzania. The genus ***Proutista*** is also well represented; the species ***P. moesta*** (Westwood) has been reported in Malaysia (Corbett 1932, Han 1978), but also by Vesey-Fitzgerald (1941) in the Seychelles and by Wilson (1987a) in Africa. It is a pest on young oil palms in Malaysia; it is parasitized by the Strepsiptera ***Halictophagous*** sp. (Ponnamma and Biju Babjan 1998). In South America, a great many species have been observed on *Elaeis*, at the adult stage only: 80 have been inventoried in Brazil (Louise 1993) and 117 in Ecuador (Perthuis 1994), mainly belonging to the genera ***Omolicna, Contigucephalus*** and ***Cedusa.***

CIXIIDAE

This family, whose larvae feed off plant roots, includes the genus ***Myndus,*** two species of which are disease vectors: ***M. crudus*** Van Duzee transmits the phytoplasma disease known as lethal yellowing, which is rife in the Caribbean (Jamaica, Cuba, Dominican Republic), Florida, Belize, Honduras and Mexico (Eden-Green 1978, Tsai 1979, Howard and Mead 1980, Howard *et al.* 1983, 1984). A positive relation has been discovered between the size of insect populations and the disease (Howard 1980). Observations were carried out on adult insects (Cherry and Howard 1984), and a study of larva host

plants was carried out (Howard 1990a, Howard and Villanueva 1994), which made it possible to envisage developing a cultural control method (Howard 1990b). Lastly, it was shown that organic mulching encouraged larva development (Howard and Oropeza 1998).
M. taffini (Bonfils 1982) transmits the virus of coconut foliar decay in Vanuatu (Julia 1982). The development cycle of this insect lasts 4 to 5 months with 6 larval instars; a female lays around a hundred eggs (Morin 1993). The larvae feed off "Bourrao" (*Hibiscus tiliaceus*) roots and the adults off coconut fronds. When *Hibiscus* is eliminated, both the insect and the disease disappear.
Lastly, another species of the genus, ***M. adiopodoumeensis*** Synave, is strongly suspected of being a lethal yellowing vector in Ghana (Dery *et al.* 1996).
Other species of the same genus have been observed on coconut: ***M. macfarlanei*** Wilson in the Santa Cruz islands, ***M. (Colvania) mavors*** Fennah (1970) in the Solomon Islands (Wilson 1988), and lastly ***M. (Colvania) rumina*** Fennah (1969) and ***M. chazeaui*** (Bourgoin and Wilson 1992) in New Caledonia.

Euryphlespia cocos Muir can cause total browning of the foliage in the Solomon Islands (Simmonds 1938).

LOPHOPIDAE

This family contains the genera ***Zophiuma,*** including ***Z. lobulata*** Ghauri involved in Finschlafen's disorder in Papua New Guinea (Smith 1980). A similar disorder has been observed in the Solomon Islands, and the species ***Painella simmondsi*** Muir has been blamed. This disease causes often reversible yellowing of the foliage, but cases of mortality can sometimes be seen (Anon. 1977).
Other genera have been mentioned: ***Virgilia and Menosca*** sp.; an outbreak of the latter species has been reported in the Philippines (Gallego 1989).

OTHER FAMILIES

In the superfamily of the *Fulgoroidea*, other species are represented in various families:
- *Ricaniidae,* with ***Ricania speculum*** Mlk on oil palm in Malaysia (Wood 1968) and ***Pochazia fasciata*** F. in the Congo (Mayné and Ghesquière 1934),
- *Meenoplidae* with several species belonging to the genera ***Anigrus, Inxwala, Kermesia*** in Sierra Leone (Hargreaves 1937),
- *Tropiduchidae* with the genera ***Numicia, Tambinia,*** including ***T. verticalis*** Distant in Zanzibar (Wilson 1987a) and others still,
- *Flatidae* eight species have been inventoried on palms (Wilson 1987a), including ***Ormenaria rufifascia*** (Walker) and ***Psenoflata brevis*** Van Duzee.

HEMIPTERA-CICADOMORPHA

CICADELLOIDEA

CICADELLIDAE

A very large number of species has been inventoried in the family of the *Cicadellidae*. The most serious among them is ***Recilia mica*** Kramer, insofar as it transmits Blast disease in oil palm seedlings in West Africa (Desmier de Chenon 1979). A species of the genus ***Nzinga*** (*Typhlocibinae*) is extremely common on oil palm and coconut fronds, where it causes discoloration of the foliage on which it spends its entire development cycle (Zebeyou *et al.* 1998). The species ***Sophonia greeni*** (Distant) is worth mentioning on young coconut palms in India (Rajan and Mathen 1984).

Studies in the Philippines, in connection with a study on *cadang-cadang* disease of coconut, led to the identification of several species: ***Anocofana mukilengensis*** (Baker), ***Sochinsogonia longa*** (Walker) and ***Bakera nigroscuta*** Mahmood (Zelazny and Pacumbaba 1982). However, these insects are not involved in transmission of the disease, which has killed over 10 million palms.

In Brazil, the palm fauna is rich in *Cicadellidae*, including ***Menosoma elegans*** (Osborn) (Louise and Filho 1993). The same applies in Ecuador, where several, often large species can reproduce on oil palm (Perthuis 1994). Such is the case with several species of the genus ***Molomea***, including ***M. fatalis*** (Bonfils and Perthuis 1992), ***Pseudometopia*** (***P. amblardii*** Signoret), ***Oncometopia*** (***O. venata*** Schröder), ***Diestostemma huallagana*** Young, etc.

OTHER FAMILIES

A few species of *Membracidae* have been inventoried in Brazil on *Elaeis,* such as ***Ceresa vitulus*** Fabricius and ***C. cuprea*** Funkhouser; these two species account for more than 65% of captures (Louise and Filho 1993).

Cercopidae belonging to the genera ***Locris, Poophilus*** and ***Clovia*** have also been observed on *Elaeis* in Côte d'Ivoire (Desmier de Chenon 1979), ***Clostoptera undulata*** Uhler on *Cocos* in Florida (Howard and Mead 1980).

Lastly, in the family of the *Cicadidae*, two species have been reported:

Dendubia intemerata Moult on coconut in Malaysia (Miller 1932) and ***Proarna hilaris*** Germar, also on coconut, in Jamaica (Eden Green and Schuiling 1973). These species lay their eggs in leaf rachises, and can cause some damage.

HEMIPTERA-STERNORHYNCHA

Within this suborder of the Hemiptera, there is a very large number of species that can be divided into three groups: whiteflies, aphids and scale insects. Whilst the number of species is large, that of pests with a true economic impact is small.

ALEYRODOIDEA

ALEYRODIDAE (whiteflies)

The insects of this family are characterized by the development of a waxy secretion, which is sometimes extremely abundant and often filiform, on the bodies of nymphs and adults. Some species coat the plant with an abundant sweet secretion on which a fungus of the genus *Aposporium* develops and can cover whole leaves, which take on a black appearance (sooty mould).

The main species is ***Aleurodicus destructor*** Mackie, which is common on coconut in Southeast Asia (Corbett 1935, Risbec 1936). Its biology and ecology have been studied in the Philippines (Sumalde and Calilung 1982, Sumalde and Gabriel 1982). Corbett (1932) reported three other species on coconut in Malaysia: ***Dialeurodes simmondsi*** Corb., ***Aleurocanthus cocois*** Corb. and ***A. gateri*** Corb. Other species have been reported (Lepesme 1947), notably in Brazil on *Cocos*: ***A. neglectus*** Quaintance and Baker, ***A. flavus*** Hempel and ***A. pulvinatus*** Maskell. In 1993/94, a severe outbreak of three species of whiteflies occurred on the island of Saint-Martin (Caribbean) including ***A. dispersus*** (Etienne and Bayart, internal report). Other infestation by ***A. cocois*** (Curtis) has been reported in the Caribbean (Martin and Watson 1998).

Many other genera of whiteflies were reported by Lepesme (1947) on *Cocos*: in Brazil with the genera ***Ceraleurodicus, Aleuronudus, Paraleyrodes,*** etc ; the genus ***Aleurocanthus*** lives off *Cocos* in Southeast Asia and the Pacific (Lepesme 1947). We observed on oil palm a severe outbreak of an undetermined whitefly in the Ecuadorian Amazon, though it was very limited in space and time.

Whiteflies can be destroyed by ladybirds, as is the case with *Megalocaria dilatata* Fab. a predator of ***Astegopteryx*** spp. and ***Aleurocantus*** spp. in Indonesia (Ginting *et al.* 1992).

APHIDOIDEA

APHIDIDAE (Aphids)

Two main species have been observed on coconut: the geographical distribution of ***Astegopteryx nipae*** Van der Goot is vast, but it is not a serious pest (Risbec 1937); the other common species are ***A. rappardi*** Lambers, ***A. rhaphidis*** Van der Goot and ***Hysteroneura setariae*** (Thomas), which are reported in the Philippines (Gabriel 1976).

Cerataphis lataniae Boisduval is also very widespread. The adults are black, edged with a waxy white fringe. In Côte d'Ivoire on *Cocos,* we found that this sweet secretion promoted abundant fumagine development. ***C. brasilensis*** (Hempel) (formerly ***C. variabilis*** Hill Ris Lambers according to Russell 1997) attacks coconut nursery plants in India (Jalaluddin and Mohanasundaran 1990) and has also been reported on *Cocos* in Suriname (Dinther 1956).

Note also ***Schzaphis rotundiventris*** (Signoret) and ***Mysteroneura setariae*** (Thomas) in India (Dhileepan 1992). We have observed ***Rhopalosiphum rufiabdominalis*** (Sasaki) on oil palm fronds in Ecuador.

COCCOIDEA (Scale insects)

These are generally highly ubiquitous and polyphagous species. A very large number of species has been observed, mostly on *Cocos*. Lepesme (1947) reported 73 species observed on *Cocos* and *Elaeis* belonging to 8 families. Since that inventory, many other species have been reported on palms, but very few species are serious pests. Ponnamma (1999) reported 20 species on oil palm in India.

DIASPIDIDAE

It is the family of the *Diaspididae* that has the largest number of species (43). It is also in this family that the most serious pests are found.

Aspidiotus destructor Signoret

Commonly known as the transparent coconut scale is one of the most serious pests on this palm. Much has been written in the literature about this species, primarily about its biological control using predator ladybirds

It occurs in virtually all countries in which coconut grows, though a few Pacific islands have remained free of it. The female lives under a waxy shield, in which the eggs are also protected in groups of around 80 on average (Mariau and Julia 1977), but fecundity is governed by climatic factors (Jalaluddin *et al.* 1992). Young first instar larvae which, with the males, are the only mobile stages, are responsible for dispersion. The female cycle lasts 4 to 6 weeks. The toxicity of the saliva causes premature senescence and necrosis of fronds. When infestation is severe, palms can be completely covered in waxy shields stuck together, causing virtually the entire coconut palm to turn yellow. The main natural enemies of ***A. destructor*** occur in various species of ladybirds including *Chilochorus nigritus* (Fabricius), *Cryptognatha nodiceps* Marshall and *Rhizobius lophanthae* (Blaisdell), *Sukunahikona popei* Vazirani (1982) in India, *Pullus xerampelinus* Muls and *Chilochorus circumdatus* Sch. in Sri Lanka, parasitized by *Aminellus indicus* Kerrich (Hym. Encyrtidae) (Sinnathamby 1977a,b), *Chilochorus schiodtei* Muls., *C. dohrni* Muls., *Cheilomenes sulfurea orbicularis* Casey, *Cheilomenes vicina* Muls. in West Africa (Mariau and Julia 1977).

Many introductions of these useful insects have been made, usually successfully. For instance, *R. lophanthae* was introduced into Vanuatu by Cochereau (1965). Thereafter, three other species were introduced on islands in the North of the country: *C. nigritus, Rhyzobius satelles* Blackburn, and *Pseudoscymnus anomalus* Chapin (Chazeau 1981). *Chilochorus cacti* L. was introduced into Sri Lanka (Sinnathamby 1977a). The Hymenoptera *Aphytis chrysomphali* Merc. (*Aphelinidae*) is a very common ectoparasite (Sinnathamby 1980), but its activity is limited by the predatory activity of ladybirds.

Lepesme (1947) mentioned three other species of ***Aspidiotus*** on *Cocos* and *Elaeis*.

Among the other species of scale insect with a degree of economic importance, we would mention ***Aonidiella cremocitri*** McKenzie, which has caused serious damage on coconut in Vanuatu (Lever 1969a; Williams and Butcher 1987, Williams and Watson 1988), ***A. orientalis*** Newstead (Fernando and Kanagaratam 1987), there was an outbreak of ***Chrysomphalus ficus*** Ashmead or "Florida red scale" in the Seychelles (Vesey-Fitzgerald 1941) and the *Aspidiotini* ***Abgrallaspis cyanophylli*** (Signoret); in Papua New Guinea, the last two species can be attacked by the ladybird of the genus *Pseudoscymnus* Chapin (Chazeau 1987). A large number of Chalcid parasitoids are known on these two species: 20 on the first and 15 on the second (Lepesme 1947).

Another species of scale insect, ***Lepidosaphes mcgregori*** Banks, has been seen infesting coconut fronds and fruits in India (Visalaksi *et* *al.* 1982), to which can also be added ***Alaucapsis*** sp. and ***Pseudolacapsis cockerelli*** (Cooley) in Sri Lanka (Fernando and Kanagarathnam 1987). Lastly, ***Furcaspis oceanica*** Lindinger attacks coconut in the Caroline Islands, where the species is controlled by *Adelencyrtus oceanicus* (Doubl), Hymenoptera *Encyrtidae* (Marutani *et al.*, 1992).

Also worth mentioning is ***Parlagena bennetti*** Bennett in Trinidad in 1969, ***Pinnaspis strachani*** Cooley on *Elaeis* fruits and ***Abgrallaspis cyanophylli*** on *Elaeis* fronds, both in Ecuador (Mariau, unpublished).

PSEUDOCOCCIDAE

This is also an important family, with 10 species of the genus ***Pseudococcus*** (Lepesme 1947), including ***P. longispinnus*** (Targioni Tozzetti) (***P. adonidum*** L.), which can attack coconut in the nursery in India (Murthy and Giridharan 1976), to which should also be added ***P. citriculus*** Green, associated with the ant *Tapinoma melanocephalum* F., ***Pseudolacapsis*** sp. in India (Jalaluddin and Mohanasundaram 1989) where ***Chrysomphalus aonidum*** (L.), ***Pseudolacapsis cockerelli*** (Cooley) and ***Aonidiella orientalis*** (Newst.) (Kapadia 1984) are found, along with ***Planococcus lilacinus*** Cockerell on coconut in Sri Lanka (Fernando and Kanagaratnam 1987) and in India (Williams and Watson 1988b). We would also mention ***Palmicultor palmarum*** (Ehrhorn) and ***Nipaecoccus nipae*** (Maskell), observed on *Elaeis* in Ecuador (Mariau, unpublished), ***Ferrisia virgata*** (Cockerell), ***Dysmicoccus cocotis*** (Maskell), ***Pseudococcus cryptus*** Hempel in the neotropical region (Williams and Granara de Willink 1992). Two species of mealybugs, ***Dysmicoccus brevipes*** Cockerell and ***Rhizoecus*** nr. ***americanus*** Hambleton, can cause damage on oil palm in Colombian nurseries (Orellana and Vera 1989).

MARGARODIDAE

Lepesme (1947) reported 4 species on oil palm and coconut, including ***Icerya seychellarum*** Westwood, considered to be the most harmful species in the Seychelles. Some species live on the roots of *Elaeis*, such as ***Dysmicoccus brevipes*** Cockerell in Ecuador (Mariau, unpublished) and ***Margarodesia desmieri*** Foldi (Foldi 1981) in West Africa. Lastly, ***Pseudaspidoproctus vaysieriellus*** Ghesquieri and ***Icerya pattersoni*** Newst have been reported on coconut in Nigeria (Aisagbonhi *et al.* 1985).

COCCIDAE

Lepesme (1947) mentioned 15 species living off *Cocos* and sometimes *Elaeis*, including ***Vinsonia stellifera*** Westwood, in the form of a starfish. Also worth mentioning in this family are: ***Gascardia*** sp. on *Elaeis* in Cameroon, ***Ceroplastes*** sp. and ***Saissetia lutea***

Newst. on coconut in Nigeria (Aisagbonhi *et al.* 1985), ***Tillancoccus*** sp. and ***Ctenochiton japonicus*** on *Elaeis* fronds in Ecuador, ***Neolecanium silverai*** Empell on *Eaeis* roots in Ecuador, where it can be found in large numbers. The economic impact of attacks is not known (Genty *et al.* 1978). ***Sangicoccus*** sp. nr. ***truncatispinus*** (Reyne) has caused substantial damage in the coconut plantings of southern Palawan, Philippines (Pacumbaba *et al.* 1997). Lastly, ***Coccus hesperidum*** (L.) on coconut in India (Fernando and Kanagaratnam 1987).

ASTEROLECANIIDAE

Several species have been identified on *Cocos nucifera* in the Peruvian Amazon: ***Palmaspis palmae*** (Cockerell), ***P. ucayali*** Matile-Ferrero, among others, on various palm species (Matile-Ferrero 1996).

HEMIPTERA-HETEROPTERA

DEPREDATORY SPECIES

These include species that attack coconut fruits, foliage and roots.

COREIDAE

This family has 4 main species: ***Amblypelta cocophaga*** China in part of the Solomon Islands (Brown 1959c), ***Pseudotheraptus wayi*** Brown in eastern Tanzania and Kenya (Way 1953a, Brown 1955, Vanderplank 1961), ***P. devastans*** Distant reported for the first time in West Africa at the beginning of the 1960s (Mariau 1969); lastly, ***Paradasynus rostratus*** Distant in India (Kurian *et al.* 1972, Raveendran Nair *et al.* 1997, Visalakshi *et al.* 1990).
The behaviour of these species is very similar; in both the larval and adult stages they puncture the flowers and nuts, which often fall within the first months of their development with species of the genus ***Pseudotheraptus***. Older nuts are found to have a lower meat content. With severe infestation, yield losses can be considerable (Vanderplank 1958a, Julia and Mariau 1978).
The main natural factor that reduces populations is the existence of an ant *Oecophylla longinoda* Latreille, which is a very active predator on the larvae of these bugs (Brown 1959ab, Vanderplank 1960, Julia and Mariau 1978, Way 1953b, 1992). If this ant is not present, nut fall can be virtually total. In Côte d'Ivoire, it has been calculated that the rate of order 4 nuts (fist size) attacked was only 10% when 80% of the palms were uniformly colonized by *Oecophylla*. Other species of ants

(*Crematogaster* sp., *Pheidole* sp., *Camponotus* sp.) have been seen to limit the speed with which *Oecophylla* becomes established (Brown 1959b, Way and Bolton 1997). Studies have been carried out on limiting *P. megacephala* populations to encourage *Oecophylla* development (Oswald 1991). In Zanzibar, with the species *P. wayi*, coconut palms with *Oecophylla* bear 57 nuts on average, and those without the ant 41 nuts (Sporleder and Rapp 1998).

Ooencyrtus utetheisae (Hymenoptera *Encyrtidae*) can parasitize the eggs of *P. wayi* (Schnauder 1993).

MIRIDAE

Carvalhoia arecae M & C, which normally lives off the areca palm, has been seen on oil palm plantlets in India. Individuals can be seriously affected by the entomopathogenic fungus *Aspergillus candidus* Link (Dhileepan *et al.* 1990).

PENTATOMIDAE

Axiagastus cambelli Distant, occurring in the western Pacific, has a similar behaviour to that of the previous four species (Lever 1969).

TINGIDAE

The adults belonging to this family are characterized by a pronotum and largely flattened hemelytra with raised lacey designs. Several species in this family puncture the leaves, but only cause slight direct damage. Such is the case with ***Stephanitis typica*** Distant, which is widespread on coconut in Asia, and which is attacked by a predatory bug *Stethoconus praefectus* Distant (Mathen and Kurian 1972). Sathiamma *et al.* (1998) inventoried 14 insects and 23 spiders that are predators of ***S. typica***. This species, for which population fluctuations are closely linked to climatic conditions (Mathen *et al.* 1968), has been suspected of being responsible for transmission of cadang-cadang disease in the Philippines (Mathen 1960), and of a coconut root disease in India (Joseph *et al.* 1972, Mathen *et al.* 1987, 1990). Another *Tingidae* has been reported in Honduras on *Elaeis*: ***Acysta interrupta*** Daly (Mexzon and Chinchilla 1991).
In Latin America, two species of *Tingidae* attack oil palm: ***Leptopharsa gibbicarina*** Froeschner (Froeschner 1976, Genty *et al.* 1975, 1983) and ***Pleseobyrsa bicincta*** Monte (Ojeda Peña et Bravo Calderon 1994). The punctures cause depigmentation of the leaflets, which can turn completely white in cases of severe infestation. However, it is not the direct damage that is the most serious. These two species provide access for a leaf fungus of the genus *Pestalotiopsis* sp. which causes small brown patches that are separate at the start of development, but eventually join up. This

parasite causes leaf desiccation and, in the event of severe infestation, entire fronds totally dry out. The species ***L. gibbicarina*** is more dangerous than ***P. bicincta*** in that it moves along the entire length of the leaflet, whereas the latter usually remains in the lower third of leaflets, a zone which plays less of a role in photosynthesis.
L gibbicarina can be affected by the fungus *Sporothrix insectorum* (Hoog. Evans) (Ordoñez and Genty 1989).

CYDNIDAE

Various species of the genus ***Scaptocoris*** have been found on palm roots in Latin America, though they do not cause any apparent damage: ***S. divergens*** in Colombia, ***S. terginus*** in Brazil, also found on alfalfa roots (Willis and Roth 1962; Brewer, 1972). Another species has also been reported on *Elaeis* roots in Brazil: ***Cirtomenus bergi*** (Moura 1983).

DISEASE VECTORS

PENTATOMIDAE

The diseases known as *Hartrot* on coconut and *Marchitez sorpresiva* on oil palm, can cause considerable damage in plantations. The etiological agent is a *Phytomonas* (flagellate protozoan of the *Trypanosomatidae* family), which develops in the sieve tubes of the phloem (Dollet and Giannotti 1976). It has been shown that various species of this family and of the genus ***Lincus*** were responsible for transmitting the disease: ***L. lethifer*** Dolling in Ecuador on *Elaeis* (Desmier de Chenon 1984, Perthuis *et al.* 1985, Segeren and Young 1985), ***L. croupius*** Rolston on *Cocos* in Guiana (Louise *et al.* 1986). Other species had already been suspected, such as ***Macropygium reticulare*** (Fabricius) (Segeren 1982). Many other species have been observed on *Cocos* or *Elaeis*: ***L. dentiger*** Fabricius in Guiana, ***L. spurcus*** Rolston and ***L. malevolus*** Rolston (Llosa *et al.* 1990), in which the glandular system has been studied (Nagnan *et al.* 1994); ***L. malevolus,*** parasitized by *Hexacladia linci* Rasplus (*Encyrtidae*) (Rasplus *et al.* 1990) in Peru, ***L. vandoesburgi*** Rolston in Suriname, ***L. tumidifrons*** in Venezuela and, lastly, ***L. lobuliger*** in Brazil (Rolston 1983, 1989, Dolling 1984). The development cycle of ***L. tumidifrons*** lasts 178 days (incubation: 8 days; 5 larval instars: 170 days) (Alvarez 1994). A structural and physico-chemical analysis of odour glands in ***L. spurcus*** has been carried out (Cassier *et al. 1994).*
A species of a neighouring genus, ***Ochlerus***, transmits the disease in Brazil (Renard 1989).

PREDATORS

Predator bugs are common in palm plantings. In both the larval and adult stages they attack a large number of leaf-eating caterpillars. They belong to two families.

REDUVIIDAE

The genus ***Sycanus*** is the most important. In Indonesia, a reared adult consumed 430 young *Metisa plana* (*Limacodidae*) caterpillars in a month (Desmier de Chenon *et al.* 1989). In this family, it is also worth mentioning ***Platymerus rhadamanthus*** Gerst, observed in Zanzibar, which attacks *Oryctes* (*Scarabaeidae)* larvae and has been introduced on several occasions under a control programme, notably in the Pacific (Vanderplank 1958b). Two other species, ***P. biguttata*** Stal. (Linnaeus) and ***Psyttala dudgeonia*** Distant have been reported in Nigeria on *Oryctes* sp. (Jerath 1965).

PENTATOMIDAE

Bugs of the family of the *Pentatomidae* play a major role in regulating populations of leaf-eating caterpillars, since they lay their eggs on palm fronds, whereas ***Sycanus*** lays its eggs in the soil. All the larval instars are predators of leaf-eating insects. In Indonesia, several species belonging to the genera ***Platynopus, Cantheconidea*** and ***Eocanthecona*** have been reported (Sipayung *et al.* 1990). Among those species, ***E. furcellata*** (Wolf) seems to be the most important. Its development cycle lasts 50 days and the female, which can reach 15 mm in length, lays 175 eggs on average. In Colombia, ***Alcorrhyncus grandis*** (Dallas) plays the same role; it has notably been observed feeding off *Limacodidae, Sibine fusca* (Stoll) (Genty 1972).

CONCLUSION

The order of the *Hemiptera* plays a very important role within the fauna of cultivated oil palm plantings in the humid Tropics. Firstly through the number of species involved, which amounts to several hundred. This order also contains some major pests, notably in the families of the *Coreidae,* such as ***Pseudotheraptus devastans*** and *Diaspididae,* such as ***Aspidiotus destructor***. Numerous viral, mycoplasma and trypanosome diseases are transmitted by Hemiptera *Fulgoromorpha* and *Cicadomorpha,* and by some *Heteroptera.* Studies are required to identify the vectors of various diseases affecting coconut and oil palm, particularly in Africa and South

America, where there is a serious lack of information. Lastly, several species of Heteroptera play an important role in regulating the populations of numerous depredators of palms, notably among the caterpillars of leaf-eating Lepidoptera.

HEMIPTERA REFERENCES

Aisagbonhi C.I., Nwana I.E., Agwu S.I. (1985). Preliminary analysis of a field population of *Aspidiotus destructor* Signoret (Homoptera: Diaspididae) and some soft-scales on coconut palms *Nigerian J. Ent.* 6 (1-2): 24-32.

Alvarez F.A. (1994). Ciclo de vida de *Lincus tumidifrons* Rolston (Hemiptera: Pentatomidae), vector de la marchitea sorpresiva de la palma de aceite. *Palmas* 15 (3): 9-16.

Anon. (1977). Annual report of the entomologist. Ministry of Agriculture and Lands Solomon Islands.

Bonfils J. (1982). Description d'une espèce nouvelle de Cixiidae nuisible aux plantations de cocotier (Homoptera Fulgoromorpha) *Bull. Soc. Ent. Fr.* 87: 381-384.

Bonfils J., Perthuis B., (1992). Liste sélective de cicadelles vivant dans une palmeraie d'Equateur et son environnement, avec description de *Molomea fatalis* n. sp. et notes écologiques. *Bull. Soc. Ent. Fr.* 97(3): 223-226.

Bourgoin Th., Wilson M.R., (1992). A new species of *Myndus* (Fulgoromorpha Cixiidae from coconut palms in New Caledonia. *Bull. Soc. Ent. Suisse* 65 (1-2): 69-74.

Brewer M. (1972.) *Scaptocoris castaneus* Perty, chinche danina a raices de alfalfa (*Hemiptera – Cydnidae*). *IDIA* 294: 27-28.

Brown E.S. (1955). *Pseudotheraptus wayi,* a new genus and species of coreid (Hemiptera) injurious to coconuts in East Africa. *Bul. Ent. Res.* 46 (1): 221-240.

Brown E.S. (1959a). Immature nutfall of coconuts in the Solomon islands. I. Distribution of nutfall in relation to that of *Amblypelta* and of certain species of ants. *Bull. Ent. Res. 59 (1): 97-133.*

Brown E.S. (1959b). Immature nutfall of coconuts in the Solomon islands. II. Changes in ant populations, and their relation to vegetation. *Bull. Ent. Res. 59* (3): 523-558.

Brown E.S. (1959c). Immature nutfall of coconuts in the Solomon islands. Notes on the life-history and biology of *Amblypelta. Bull. Ent. Res. 59* (3): 559-566.

Cassier P., Nagnan P., Llosa J.F., Andre M., Guillaumin D. (1994). Fine structure and physicochemical analysis of the nymphal and imaginal scent gland systems of *Lincus spurcus* Rolston (Heteroptera: Pentatomidae). *Int. J. Insect Morphol. & Embryol.* 23 (1): 1-12.

Chazeau J. (1981). La lutte biologique contre la cochenille transparente du cocotier *Temnaspidiotus destructor* (Signoret) aux Nouvelles-Hébrides (Homoptère *Diaspididae). Cahier O.R.S.T.O.M., série Biol.* 44: 11-22.

Chazeau J. (1987). Complément aux *Scymnus* et *Nephlus* de Nouvelle-Guinée et des archipels voisins. *Bull. Soc. Ent. Fr.* 92 (1-2): 23-28.

Cherry R.H., Howard F.W. (1984). Sampling for adults of the plant hopper *Myndus crudus,* a vector of lethal yellowing of palms. *Tropical Pest Management* 30 (1): 22-25.

Cochereau P. (1965). Contrôle biologique d'*Aspidiotus destructor* Sign. dans l'île Vaté, Nouvelles Hébrides, au moyen de *Lindorus lophanthae* Blaisd. *C.R. Acad. Agric. France* 51(5): 318-321.

Corbett G.H. (1932). Insects of coconut in Malaya. *Dept. Agric. S.S.F.M.S. Gen. Ser.* 10: 1-106.

Corbett G.H. (1935). Malayan *Aleyrodidae. Journal F.M.S. Mus.* XVII: 722.

Dery K., Philippe R., Mariau D. (1996). *Auchenorryncha* (*Homoptera*). Suspected vectors of Lethal Yellowing disease in Ghana. *Plantation Recherche Développement* 3 (5): 355-363.

Desmier de Chenon R. (1979). Mise en évidence du rôle de *Recilia mica* Kramer (*Homoptera, Cicadellidae, Deltocephalinae*) dans la maladie du Blast des pépinières de palmier à huile en Côte d'Ivoire. *Oléagineux* 34 (3): 107-112.

Desmier de Chenon (1984). Recherches sur le genre *Lincus* Stål, *Hemiptera Pentatomidae Discocephalinae,* et son rôle dans la transmission de la Marchitez du palmier à huile et du Hartrot du cocotier. *Oléagineux* 39 (1): 1-4.

Desmier de Chenon R., Sipayung A., Sudharto P.S. (1989). The importance of natural enemies on leaf-eating caterpillars oil palm plantations in Sumatra, Indonesia. Uses and possibilities. *PORIM, International Palm Oil Development Conf.*, 5-9 Sept. 1989, 26 p. + annexes.

Dhileepan K. (1992). Insect pests of oil palm (*Elaeis guineensis)* In India. *Planter* 68: 183-191.

Dhileepan K., Nair R.R., Leena S. (1990). *Aspergillus candidus* Link as an entomopathogen of spindle bug *Carvalhoia arecae* M & C (Miridae: Heteroptera). *Planter* 66: 519-521.

Dinther van J.B.M. (1956). Insects of the coconut palm in Suriname. *Landbouwproefstation in Suriname Bulletin* N° 69, 27 p.

Dollet M., Giannotti JJ. (1976). Maladie de Kainkopé: présence de mycoplasmes dans le phloème des cocotiers malades. *Oléagineux* 31 (4): 169-175.

Dolling W.R. (1984). Pentatomid bugs (Hemiptera) that transmit a flagellate disease of cultivated palms in South America. *Bull. Entomol. Res.* 74: 473-476.

Eden-Green S.J. (1978). Rearing and transmission techniques for *Haplaxius* sp. (Hom: Cixiidae), a suspected vector of lethal yellowing disease of coconuts. *Ann. appl. Biol.* 89: 173-176.

Eden-Green S., Schuiling M. (1973). Some attempts to rear leaf hopper vectors of lethal yellowing of coconut palm in Jamaica. *Plant Disease Reporter* 62: 625-627.

Eskafi F.M. (1982). Leafhoppers and planthoppers feeding on coconut in Jamaica. *Trop. Agric.* 54 (4): 289-292.

Fennah R.G. (1964). *Delphacidae* from Madagascar and the Mascarenes (*Homoptera*: *Fulgoroidea*). *Transactions of the Royal Entomological Society of London* 116: 131-150.

Fennah R.G. (1969). Fulgoroidea (Homoptera) from New Caledonia and the Loyalty islands. *Pac. Insects Monogr.* 21: 1-116.

Fennah R.G. (1970). Fulgoroidea (Homoptera) from Rennell and Bellona islands. *The Natural History of Rennell Island* 6: 43-85.

Fernando L.C.P., Kanagaratnam P. (1987). New records of some pests of the coconut inflorescence and developing fruit and there natural enemies in Sri Lanka. *Cocos* 5 (3): 9-42.

Foldi I. (1981). *Margarodesia* N. Gen. Cochenille afro-tropicale nuisible au palmier à huile (*Homoptera, Coccoidea, Margarodidae*). *Annls.. Soc. Ent. Fr.* 17 (3): 331-339.

Froeschner R.C. (1976). Description of new species of lace bug attacking the oil palm in Colombia (Hemiptera Tingidae). *Proc. Ent. Soc. Washington* 78 (1): 104-107.

Gabriel B.P. (1976). Pests of coconut in the Philippines. *Philippines J. Coconut Studies* 1 (1): 15-26.

Gallego V.C. (1989). Outbreak of *Menosca* sp. (Homoptera: Lophopidae) on coconuts in Valencia, Negros oriental. *Philippines J. Coconut Studies* 14 (1): 7-9.

Genty P. (1972). Morphologie et biologie de *Sibine fusca* Stoll, lépidoptère défoliateur du palmier à huile en Colombie. *Oléagineux* 27 (2): 65-71.

Genty P., Desmier de Chenon R., Morin J.P., Korytkowoski C.A. (1978). Ravageurs du palmier à huile en Amérique Latine. *Oléagineux* 33 (7): 325-419.

Genty P., Garzon A., Garia R. (1983). Dégâts et contrôle du complexe *Leptopharsa – Pestalotiopsis* chez le palmier à huile. *Oléagineux* 38 (5): 291-299.

Genty P., Gildardo Lopez J., Mariau D., 1975. Dégâts de *Pestalotiopsis* induits par les attaques de *Gargaphia* en Colombie. *Oléagineux* 30(5): 199-204.

Ginting C.U., Djamin A., Lamria Sidaruk (1992). Some biological aspects of *Megalocaria dilatata* Fab. (Col: Coccinellidae) a potential coconut leaf aphid predator. *Oléagineux* 47 (10): 569-572.

Han K.J. (1978). *Proutista moesta* a pest of young oil palms. *Planter* 54 (630): 522-526.

Hargreaves E. (1937). Some insects and their food plants in Sierra Leone. *Bull. Ent. Res.* 28: 505-520.

Howard F.W. (1980). Population densities of *Myndus crudus* Van Duzee (Homoptera: Cixiidae) in relation to coconut lethal yellowing distribution in Florida. *Principes* 24 (4): 174-178.

Howard F.W. (1990a).Evaluation of *Brachiaria* and St. Augustin grasses as breeding hosts of *Myndus crudus* Van Duzee (*Homoptera: Cixiidae)*, a vector of the lethal yellowing disease of palms. *Beitr. Trop. Landwirtsch. Vet. Med.* 28 (H. 4): 425-430.

Howard F.W. (1990b). Evaluation of grasses for cultural control of *Myndus crudus,* a vector of lethal yellowing of palms. *Entomol. exp. appl.* 56 (2): 131-137.

Howard F.W., Mead F.W. (1980). A survey of *Auchenorryncha* (Insecta: Homoptera) associated with palms in Southern Florida. *Trop. Agric.* 57(2): 145-153.

Howard F.W., Norris R.C., Thomas D.L. (1983). Evidence of transmission of palm lethal yellowing agent by a planthopper, *Myndus crudus* (Homoptera: Cixiidae). *Trop. Agric.* 60 (3): 168-171.

Howard F.W., Oropeza C. (1998). Organic mulch a factor in the nymphal habitat of *Myndus crudus* (Hemiptera: Auchenorryncha: Cixiidae). *Florida Entomologist* 81 (1): 92-97.

Howard F.W., Villanueva B. (1994). The abundance of the American palm cixiid in relation to grass height and location of nymphs in relation to soil surface. *J. Appl. Ent.* 118: 239-244.

Howard F.W., Williams D.S., Norris R.C. (1984). Insect transmission of lethal yellowing to young palms. *International Journal of Entomology* 26 (4): 331-338.

Jalaludin S.M., Mohanasundaram M. (1989). New records and distribution of coconut coccids in Tamil Nadu, India. *Indian Coconut J.* 10 (1): 11-14.

Jalaluddin S.M., Mohanasundaram M. (1990). Efficacy of insecticides against coconut aphid *Cerataphis variabilis* in the nursery. *Indian Coconut J.* 11 (1): 16-17.

Jalaluddin S.M., Mohanasundaram M., Sundarababu (1992). Influence of weather on fecundity potential of coconut scale *Aspidiotus destructor* Sign. *Indian Coconut J. 22: 5-7.*

Jerath M.L., 1965. A list of insects found on palms in Nigeria and their known parasites and predators. *Jour. of Nifor* 1965: 411-415.

Joseph T., Shanta P., Lal S.B. (1972). Role of *Stephanitis typicus* Distant in the spread of coconut root (wilt) pathogen. *Indian J. Agric. Sci.* 42 (5): 414-417.

Julia J.F. (1982). *Myndus taffini* (Homoptera Cixiidae), vecteur du dépérissement foliaire des cocotiers au Vanuatu. *Oléagineux* 37(8-9): 409-414.

Julia J.F., Mariau D. (1978). La punaise du cocotier, *Pseudotheraptus p.*, en Côte d'Ivoire. I. Etudes préalables à la mise au point d'une méthode de lutte intégrée. *Oléagineux* 33 (2): 65-75.

Julia J.F., Mariau D. (1982). Deux espèces de *Sogatella* (Homoptère *Delphacidae*), vectrices de la maladie de la pourriture sèche du coeur des jeunes cocotiers en Côte d'Ivoire. *Oléagineux* 37(11): 517-520.

Kapadia M.N. (1984). New records of coconut scale insects and their enemies in Gujarat.*Indian Coconut J.* 5: 9-10.

Kurian C., Pillai G.B., Abraham V.A., Matheu K. (1972). Record of a coreid bug (nut crinkler) as a new pest of coconut in India. *Curr. Sci.* 41 (1): 37.

Lepesme P. (1947). Les insectes des palmiers. Paul Lechevalier ed., Paris, 903 p.

Lever R.J.A.W. (1969). Les ravageurs du cocotier. *Etudes agricoles de la FAO*, no 77, Rome, Italie, 196 p.

Llosa J.F., Couturier G., Khan F. (1990). Notes on the ecology of *Lincus spurcus* and *L. malevolus* (Heteroptera: Pentatomidae: Discocephalinae) on Palmae in forests of peruvian Amazonia. *Ann. Soc. Ent. Fr.* (N.S.) 26 (2): 249-254.

Louise C., Dollet M., Mariau D. (1986). Recherches sur le hartrot du cocotier, maladie à *Phytomonas* (*Tripanosomatidae*) et sur son vecteur *Lincus* sp. (*Pentatomidae*) en Guyane. *Oléagineux* 41(10): 437-449.

Louise C., Filho C.P. (1993). Recherches d'un insecte vecteur de l'amarelecimento fatal ("AF") sur une plantation de palmier à huile du Nord du Brésil. Rapport Cirad, 12 p.

Mariau D. (1969). *Pseudotheraptus:* un nouveau ravageur du cocotier en Afrique occidentale. *Oléagineux* 24 (1): 21-25.

Mariau D. (1998). Hemiptera (Insecta) on oil palm and coconut. *Insect Sci. Applic.* 18 (4): 269-277.

Mariau D., Julia J.F. (1977). Nouvelles recherches sur la cochenille du cocotier *Aspidiotus destructor* Signoret. *Oléagineux* 32(5): 217-224.

Marutani M., Muniappan R., Denton G.R.W. (1992). Biological Control of red Coconut scale in Guam. *Journal of Plant Protection in the Tropics* 9 (1): 63-67.

Martin J.H., Watson G.W. (1998). *Aleurodicus pulvinatus* (Maskell) (Homoptera: Aleyrodidae), ant its emergence as a pest of coconut in the Caribbean. *Journal of Natural History* 32: 85-98.

Mathen K. (1960). Observations on *Stephanitis typicus*, a pest of coconut palms. I. Description and life history. *Ind. Coconut J.* 14(1): 8-27.

Mathen K., Kurian C. (1972). Description, life-history and habits of *Stethoconus praefectus* Distant (Heteroptera: Miridae), predacious on *Stephanitis typicus* Distant (Heteroptera: Tingidae), a pest of coconut palm. *Indian J. agric. Sci. 42* (3): 255-262.

Mathen K., Mathew J., Kirian C. (1968). Seasonal abundance of the population of *Stephanitis typicus* Distant in the field of coconut palm. *Indian J. agric. Sci.* 38 (4): 644-653.

Mathen K., Solomon J.J., Rajan P., Geetha L. (1987). Electron microscopic evidence on the role of *Stephanitiis typica* (Distant) as vector of coconut root (wilt) disease. *Current Science* 56 (23): 1239-1241.

Mathen K., Rajan P., Radhakrishnan Nair C.P., Sasikala M., Gunasekharan M., Govindankutty M.P., Solomon J.J. (1990). Transmission of root (wilt) disease to coconut seedlings through *Stephanicus typica* (Distant) (Heteroptera: Tingidae). *Trop. Agric.* 67 (1): 69-73.

Matile-Ferrero D. (1996). La faune des Asterolecaniidae palmicoles de l'Amazonie péruvienne (Homoptera: Coccoidea). *Ann. Soc. Ent. Fr.* 32 (3): 251-264.

Mayne R., Ghesquiere J. (1934). Hémiptères nuisibles aux végétaux du Congo Belge. *Ann. Gembloux* XL: 1-41.

Mexzón R.G., Chinchilla C.M. (1991). Entomofauna prejudicial, enemigos naturales y malezas utiles en Palma aceitera (*Elaeis guineensis*). J. en America Central. *Manejo Integrado de Plagas* (Costa Rica) 20-21: 1-7.

Miller N.C.E. (1932). Preliminary list of food-plants of some Malayan insects. Suppl. *Bull. Dept. Agr.* F.M.S.

Morin J.P. (1993). Elevage et biologie de *Myndus taffini* Bonfils (Homoptera: Cixiidae) vecteur du dépérissement foliaire du cocotier au Vanuatu. ANPP troisième conférence internationale sur les ravageurs en agriculture. Montpellier 7-8-9 décembre 1993: 1375.

Moura J.I.L. (1983). Danos causados ao dende (*Elaeis guineensis*) por acao do *Cirtomenus bergi* (Froeschner 1960) (Hemiptera Cydniidae). *Revista Floresta* XIV (2): 59-60.

Murthy A.D., Giridharan S. (1976). Control of the coconut mealy bug. *Pseudococcus longispinus* T. *Coconut bulletin* 6 (12): 3.

Nagnan P., Cassier P., André M., Llosa J.F., Guillaume D. (1994). Fine structure and physicochemical analysis of metathoracic glands of *Lincus malevolus* (Rolston) and *L. spurcus* (Rolston) (Heteroptera: Pentatomidae). *Int. J. Insect Morphol. & Embryol.* 23 (1), 16 p.

Orellana F., Vera H.D. (1989). Las "Cochinillas harinosas" *(Dysmicoccus brevipes* Cockerell y *Rhizoecus* prob. *americanus.* Hamilleton) en viveros de palma africana y su control. *INIAP Ecuador, Boletin Divulgativo* n° 200, 8 p.

Ojeda Peña D., Bravo Calderon M. (1994). Biologia y control de *Pleseobyrsa bicincta* Monte (Hem.: Tingidae), plaga importante de palma aceitera en el Perú. *Oléagineux* 49 (4): 145-154.

Ordoñez A.I., Genty P. (1989). Evaluacion del hongo *Sporothrix insectorum* (Hoog, Evans) en la chinche de encaje *Leptopharsa gibbicarina* (Froeschner) en palma africana de aceite *(Elaeis guineensis* Jacq.). *Revista colombiana de Entomologia* 15 (2): 10-20.

Oswald S. (1991). Application of the selective fire and bait AMDRO against the harmful brown ant, *Pheidole megacephala,* for improvement of the biological control of the coconut bug, *Pseudotheraptus wayi,* by the beneficial ed ant, *Oecophylla longinoda. Journal of Plant Diseases and Protection* 98 (4): 358-363.

Pacumbaba E.P., Zipagan M.B., Orense J.C., Imperial M.L.R. (1997). *Sangicoccus* sp. nr. *truncatispinus* (Reyne) a mealybug species infesting coconut palms in southern Palawan. *Philippine J. Crop Science* 22.

Perthuis B. (1994). Recherches d'insectes vecteurs de la pourriture du coeur du palmier à huile (*Elaeis guineensis* Jacq.) en Equateur oriental: sélection des homoptères *Auchenorrhyncha* les plus intéressants à tester. Rapport Cirad, 34 p.

Perthuis B., Desmier de Chenon R., Merland E. (1985). Mise en évidence du vecteur de la Marchitez sorpresiva du palmier à huile, la punaise *Lincus lethifer* Dolling, *Hemiptera Pentatomidae Discocephalinae. Oléagineux* 40 (10): 473-476.

Ponnamma K.N. (1999). Coccoids associated with oil palm in India, a review. *Planter* 882: 445-451.

Ponnamma K.N., Biju Babjan. (1998). Record of *Halictophagous* sp. (Strepsiptera: Halictophagidae), a parasitoid of *Proutista moesta* (Westwood) (Homoptera: Derbidae), a vector of phytoplasma diseases of palms. *Planter* 74 (862): 37-40.

Rajan P., Mathen K. (1984). *Sophonia greeni* (Distant) (Nirvanidae: Jassoidea) on leaves of Coconut palm *Cocos nucifera* L. *Journal of Plantation Crops* 12 (2): 178-179.

Rasplus J.Y., Pluot-Sigwalt D., Llosa J.F., Couturier G. (1990). *Hexacladia linci* N.sp. (Hymenoptera Encyrtidae) endoparasite de *Lincus malevolus* Rolston (Heteroptera: Pentatomidae) au Pérou. *Ann. Soc. Ent. Fr.* (N.S.) 26 (2): 255-263.

Raveendran Nair K.K., Visalakshi A., Nandakumar C., Naseema Beevi S., Reghunath P., Sasidharan Pillai K. (1997). Quantitative loss due to nut crinkler (*Paradasymus rostratus* Dist). Damage on coconut. *Entomon.* 22 (3-4): 239-241.

Renard J.L. (1989). Le hartrot du cocotier. Caractérisation et moyens de lutte. *Oléagineux* 44 (10): 475-482.

Risbec J. (1936). Les parasites du cocotier en Nouvelle Calédonie. *Agronomie coloniale* 222: 173-178.

Risbec J. (1937). Les parasites des plantes cultivées aux Nouvelles Hébrides. *Faune des colonies françaises* T-VI: 76-77.

Rolston L.H. (1983). A revision of the genus *Lincus* Stäl (Hemiptera Pentatomidae: Discocephalinae: Ochlerini). *New York entomological Society* 91(1): 1-47.

Rolston L.H. (1989). Three new species of *Lincus* (Hemiptera: Pentatomidae) from palms. *J. New York Entomol. Soc.* 97 (3): 271-276.

Russel L.M. (1977). Notes on *Cerataphis brasiliensis* and synonyms *palmae, variabilis* and *fransseni* (Homoptera:Aphididae), with a key to *Cerataphis* species living on palms and orchids. *Proceedings of the Entomological Society of Washington* 98: 439-449.

Sathiamma B., Chandramohanan K.R., Soniya V.P. (1998). Record of the natural enemies of the lace bug *Stephanitis typica* (Distant) a pest on coconut palm. *Entomon.* 23 (4): 321-324.

Schnauder C. (1993). Eiparasitismus bei *Pseudotheraptus wayi* Brown (Heteroptera, Coreidae) durch die schlufwespe *Ooencyrtus utetheisae (*Risbec) (Hymenoptera, Encyrtidae). *Mitt. Dtsch. Ges. Allg. Angew. Ent.* 8: 821-825.

Segeren P. (1982). Preliminary study on the vector (s) of hartrot disease of coconut in Suriname. *De Surinaamse Landbouw* 30 (1): 17-23.

Segeren P., Young R. de (1985). Inventory of possible insect vectors of hartrot disease of coconut in Suriname. *De Surinaamse Landbouw* 33 (1): 21-28.

Sinnathamby S.V. (1980). The behaviour of *Aphytis chrysomphali,* a hymenopterus parasie of the coconut scale, *Aspidiotus destructor,* in Sri Lanka. *Ceylon Coconut Q.* 31: 93-97.

Sipayung A., Desmier de Chenon R., Sudharto P. (1990). Study of the *Eocanthecona, Cantheconidea* predator complex in Indonesia (*Hemiptera, Pentatomidae, Asopinae*), 3rd Int. Conf. on Plant Protec. in the Tropics. Malaysian Plant Protection Society, Genting Highlands Pahang, Malaysia 20-23 March 1990, 32 p.

Simmonds H.W. (1938). Coconut pests and diseases in Melanesia and Southern Polynesia.*Departement of Agriculture Bull.* N° 20, 39 p.

Sinnathamby S.V. (1977a). Developments in the control of coconut scale, *Aspidiotus destructor* Sign. in Sri Lanka. *Ceylon Cocon. Q.* 28: 81-88.

Sinnathamby S.V. (1977b). Studies on the biology of *Pullus xerampelinus* Muls. (Coleoptera: Coccinellidae), a predator of *Aspidiotus destructor* Sign. (Homoptera: Coccidae), with a note on its parasite. *Ceylon Cocon. Q.* 28: 94-100.

Smith E.S.C. (1980). *Zophiuma lobulata* Ghauri (*Homoptera*: *Lophopidae*) and its relation to the Finschlafen coconut disorder in Papua New Guinea. *Papua New Guinea Agricultural Journal* 31: 37-45.

Sporleder M., Rapp G. (1998). The effect of *Oecophylla longinoda* (Latr.) (Hym., Formicidae) on coconut palm productivity with respect to *Pseudotheraptus wayi* Brown (Hem., Coreidae) damage in Zanzibar. *J. Appl. Ent.* 122: 475-481.

Sumalde A.C., Calilung V.J. (1982). Life history of the aphid *Astegopteryx nipae* (Van der Goot) (Pemphigidae, Homoptera) on coconut (*Cocos nucifera L.*) *Philippine Journal of Coconut Studies* VII (1-2): 18-22.

Sumalde A.C., Gabriel B.P. (1982). Biology and ecology of the coconut white fly (*Aleurodicus destructor* Mackie) (Aleyrodidae Homoptera). *Philippine Journal of Coconut Studies* VII (1-2): 9-17.

Tsai J.F. (1979). Vector transmission of mycoplasmal agents of plant diseases *In:* The mycoplasmas, vol. III R.F. Whitcomb et J.G. Tully ed., New York. Etats-Unis: 266-307.

Tsai J.H., Fisher J.B.(1993). Feeding sites of some leaf and planthopper insects (Homoptera: Auchenorrhyncha). Associated with coconut palms. *Principes* 37 (1): 35-41.

Vanderplank F.L. (1958a). Studies on the coconut pest, *Pseudotheraptus wayi* Brown (Coreidae), in Zanzibar. I. Method of assessing the damage caused by the insect. *Bull. Ent. Res.* 49 (3): 559-584.

Vanderplank F.L. (1958b). Studies on the coconut bug, *Platymerus rhadamanthus* Gerst. a useful predator of *Oryctes boas* and *O. rhinoceros*. *J. Ent. Soc. S. Africa*. II 21(2): 309-314.

Vanderplank F.L. (1960). The bionomics and ecology of the red tree ant, *Oecophylla* sp., and its relationship to the coconut bug *Pseudotheraptus wayi* Brown (Coreidae). *J. Anim. Ecol.* 29 (1): 15-33.

Vanderplank F.L. (1961).Availability of the coconut bug *Pseudotheraptus wayi* (Coreidae). *Bull. Ent. Res.* 51: 57-60.

Vazirani T.G. (1982). *Sukunahikona popei* sp. n. (Coleoptera: Coccinellidae) feeding on scale insects infesting coconut palm in Gujarat, India. *Bull. Ent. Res.* 72: 29-32.

Vesey-Fitzgerald D. (1941). Progress of the control of coconut feeding *Coccidae* in Seychelles. *Bull. Ent. Res.* 32, p. 161.

Visalakshi A., Ravindran Nair K.K., Nalinakumari T. (1982). *Lepidosaphes Mcgregori* Banks a new hard scale infesting coconut. *Entomon.* 7 (4): 500.

Visalakshi A., Nandakumar C., Reghunath P., Mohandas N. (1990). Occurrence of *Paradasynus rostratus* Dist (Coreidae) as a menace to coconut plantations in Kerala (S. India) and management of the pest. 3rd Int. Conf. on Plant Protec. *In* the Tropics. 6: 175-179.

Way M.J. (1953a). Studies of *Theraptus* sp. (Coreiddae), The cause of the gumming disease of coconuts in East Africa. *Bull. Ent. Res.* 44: 657-667.

Way M.J. (1953b). Studies of the life history and ecology of the ant *Oecophylla longinoda* (Latreille). *Bull. Ent. Res.* 45: 93-112.

Way M.J. (1992). Role of ants in Pest management. *Ann. Rev. Entomol.* 37: 479-503.

Way M.J., Bolton B. (1997). Competition between ants for coconut palm nesting sites. *Journal of Natural History* 31: 439-455.

Williams D.J., Butcher C.F. (1987). Scale insects (Hemiptera: Coccoidea) of Vanuatu. New Zealand Entomologist, 9: 88-99.

Williams D.J., Watson G.W. (1988a). The scale insects of the tropical South Pacific region. Part I. The armoured scales (Diaspididae). CAB International, Wallingford, UK., 290 p.

Williams D.J., Watson G.W. (1988b). The scale insects of the tropical South Pacific region. Part 2. The mealybugs (Pseudococcidae). CAB International, Wallingford, UK., 260 p.

Williams D.J., Granara de Willink M.C. (1992). Mealybugs of Central and South America. CAB International, Wallingford, UK., 635 p.

Willis E.R., Roth L.M.. (1962). Soil and moisture relations of *Scaptocoris* divergens Froeschner (Hemiptera: Cydnidae). *Annuals of the Ent. Soc. of America* 55: 21-33.

Wilson M.R. (1987a). The *Auchenorryncha* (*Homoptera*) associated with palms. Proc. 2nd Int. Workshop of Leafhoppers and Planthoppers of Economic Importance, held Provo, Utah, USA, 28th July, 1st Aug. 1986. Eds M.R. Wilson G.L.R. Nault, CIE, London, 1987: 327-342.

Wilson M.R. (1987b). African Derbidae (*Homoptera, Fulgoridea*): Taxonomic notes with descriptions of new species collected mainly from coconut. *J. Nat. History* 21: 567-595.

Wilson M.R. (1988).The genus *Myndus* (Hemiptera: Cixiidae) in the Solomon islands and Vanuatu and its relation to foliar decay of coconut palms in Vanuatu. *Bull. Ent. Res.* 78: 519-526.

Wood B.J. (1968). Pests of oil palms in Malaysia and their control. The incorporated society of Planters, Kuala Lumpur, Malaysia, 204 p.

Zebeyou M.G., Philippe R., Mariau D. (1998). The description and biology of *Nzinga palmivora*, (*Hemiptera, Cicadellidae, Typhlocibinae*) from cultivated palms in pest in West Africa (rapport non publié).

Zelazny B., Pacumbaba E. (1982). Phytophagous insects asociated with Cadang-Cadang infected and healthy coconut palms in Southern-Eastern Luzon, Philippines. *Ecological entomology 7*: 113-120.

COLEOPTERA

INTRODUCTION

Many species, mostly belonging to the family of the *Scarabaeidae*, attack oil palm and coconut in their adult form, whereas for other species, particularly of the families *Curculionidae* and *Chrysomelidae,* it is the larval instars that cause damage. A large number of them figure among the most serious palm predators.

However, many Coleoptera species are only linked to these plants because their larvae feed off rotting stems. Such is the case for insects belonging to the families of the *Lucanidae, Passalidae*, certain *Elateridae, Tenebrionidae, Cerambycidae,* etc, or to the subfamily of the *Cetoniinae*, even though the adults of a certain number of species in this subfamily, such as the rose beetles, feed off pollen, or can cause some minimal damage to foliage.

Apart from the odd exception, the insects of those families will not be covered here, as they cannot really be considered as true palm pests. However, we would mention several species of the genus ***Eurytrachelus*** *(Lucanidae).* Some adults of a few species in this family can penetrate the stem and leaf axils in the Solomon Islands and in Papua New Guinea (Lever, 1969). Likewise, ***Mulciber linnaei*** *(Cerambycidae)* has been known to cause substantial damage to 2-year-old palms in Papua New Guinea (OPRA, 1988) (see other Coleoptera section).

The main pests primarily belong to three families, the *Scarabaeidae, Chrysomelidae* and *Curculionidae*. A few other pests, which are sometimes locally serious, are classed in families of lesser importance. Coleoptera in stored foodstuffs will be dealt with separately.

SCARABAEIDAE

RUTELINAE

The larvae of this subfamily feed off a great variety of roots, notably those of herbaceous plants at the foot of palms. The genus ***Adoretus*** includes several species. In Asia, ***A. compressus*** Webb and ***A. borneensis*** Kraatz, whose adults are 1.5 cm long, have a light brown body covered with fine white hairs. The adults, which are nocturnal, attack the fronds of young palms and can cause a fair amount of damage (Mariau *et al.* 1991). In India, we would mention ***A. lasiopygus*** Burmeister (Johnston 1965) and ***A. lithobius*** Ohans (Pillai 1921). In Asia, Lepesme (1947) mentioned ***A. sinicus*** Burm., ***A. dorsalis*** F. and ***Macraspis tetradactyla*** L. in Jamaica. In Africa, ***A. umbrosus*** Fabricius plays the same role (Mariau *et al.* 1981), as well as ***A. celogaster*** Arz in Sri Lanka (FAO 1964).

MELOLONTHINAE

Some species have a very similar behaviour to that of ***Adoretus***. In Asia, we would mention ***Leucopholis coneophora*** Burmeister whose larvae attack the roots of *Cocos* in India in the State of Kerala (Nirula *et al.* 1952). It is considered to be a serious pest in the States of Kerala and Karnataka; the development cycle of this species lasts a little over 10 months (Abraham and Mohandas 1988). ***Psilopholis vestita*** also attacks the roots of young palms in Malaysia (Wood and Kwang Yew 1969). Also worth mentioning are ***Apogonia expeditionis*** Ritsema, which has a smooth dark brown body measuring approximately 1 cm (Mariau *et al.,* 1991) and ***Sophrops eurystoma*** Burmeister in Sri Lanka. In Africa, there are ***Pseudotrochatus concolor*** Kolbe, ***P. nigroviridis*** (Ghesquière *in* Burgeon *in* Lepesme 1947) and ***Schizonycha africana*** Castelnau, which has a 15 mm long reddish-brown body, with white spots (Mariau *et al.* 1981).
Sexual pheromones have been discovered in several species of *Melolonthinae* (Leal 1998).

TRICHIINAE

A few species live in Africa at the base of fronds, mining galleries in the underside of the rachis. The most common species is ***Platygenia barbata*** Afzal (Alibert 1938, Frappa 1933). It is very common in palm crowns. More than a dozen adults can be found together in a single crown, though damage appears to be secondary (Mariau, unpublished). The larvae can be attacked by the carabid beetle *Neochryopus savagei* Hope (Lepesme 1947). ***Incala setosus*** Mos. shows a similar behaviour (Ghesquière 1935), as does ***Agenius*** (***Clastocnemis***) ***quadrimaculatus*** Afzel, which can also be found on nursery seedlings (Alibert 1938), and ***A. colini*** Mosec (Magné *in* Lepesme 1947).

CETONIINAE

We shall not be mentioning the very large number of species found in their adult form on *Elaeis* and *Cocos* male flowers, where they feed off the pollen, and whose larvae live in rotting wood. They are not true palm pests. However, ***Pachnoda marginella*** F. has been reported to mine larva galleries in rachises and even in the stems of palms (Ghesquière 1935 *in* Lepesme 1947).

DYNASTINAE

This subfamily includes serious pests belonging to the main genera ***Oryctes, Scapanes*** and ***Strategus.***

Oryctes[•]

Given the economic importance of these insects, the literature is particularly abundant. In work on ***O. rhinoceros*** L., dating back almost fifty years, there were already 310 references (Gressitt 1953). Twenty years ago, 191 references were listed by Bedford (1980). However, only a few species are serious pests: ***O. rhinoceos,*** which is widespread in Asia, throughout most of the islands of the Pacific Ocean, and more recently in Mauritius (Vinson 1963) and Réunion (Monty 1978). ***O. monoceros*** Olivier is found in West and East Africa, and in the Seychelles. ***O. boas*** Fabricius is African and was introduced into Madagascar, where 11 species are endemic, though two species can cause substantial damage on coconut there: ***O. simiar*** Coquerel and ***O. pyrrhus*** Burmeister. Many other species are reported on palms, especially *Cocos,* but have never been seen to be serious pests. Of these species, we would mention ***O. augias*** Olivier, ***O. chevrolati*** Guérin, ***O. insularis*** Coquerel in Madagascar, ***O. gigas*** Castelnau and ***O. erebus*** Burmeister in Africa, ***O. centaurus*** Sternbery in Papua New Guinea, ***O. comoriensis*** Fairmaire in the Comoro Islands, along with many others.

Description

The adult is dark brown. The adult of ***O. monoceros*** and ***O. rhinoceros*** is around 4 cm long. The head bears a short horn, which is more developed in the male and in larger individuals. They can be much larger in size, up to 70 mm for ***O. gigas*** and ***O. centaurus,*** which possess a powerful horn.

The larva is a typical white grub, with a thick, cylindrical and highly curved body. The head is highly sclerotized and has strong mandibles. Although very similar from one species to the next, several of the *Scarabaeidae* species can be distinguished between in their larval form through their degree of pilosity (Bedford 1974, Beaudoin-Ollivier *et al.* 1998).

Biology

All these species have a very similar biology and behaviour. The larvae live in rotting plant matter, with a preference for palm stems.

The larvae of ***O. boas*** have only been found in compost in Africa, limiting attacks by that species to coconut palms near villages (Mariau 1967).
The development cycle lasts 4 to 5 months depending on the species, including 3 to 4 for the larval cycle, which involves 3 instars, and around 3 weeks for pupation, which takes place in a fibrous case.
In the case of ***O. monoceros***, the adults mine a gallery at the base of unopened fronds, using the lower section of rachis on frond ranks 1 to 3 for support. On adult palms, they have been seen to mine into the rachis of very young fronds, which then snap off in the wind (Mariau, unpublished). In the first year after planting, young coconut palms can be killed by an attack or emit severely deformed fronds. Yield losses of up to 40% in the first harvesting year have been reported in Malaysia on oil palm (Fee *et al.* 1989). The adults of this species can fly for a distance of several hundred metres, but most of them only cover short distances (Mariau 1970). Attacks may occur in fronds of lower ranks in the case of ***O. rhinoceros*** and much lower for ***O. simiar***, making these last two species more dangerous, since the adults reach the terminal bud more easily. The feeding gallery of ***O. monoceros*** can reach 1 metre in length and the adult mines about 4 cm per day (Mariau 1968). ***O. pyrrhus*** only attacks seedlings, digging a hole in the soil a few centimetres from the plant, then penetrating the coconut palm after passing through the husk (Mariau, unpublished). Its behaviour is very similar to that of ***Strategus*** (see below).
In the first year after planting, young coconut palms can be killed by an attack or emit severely deformed fronds. Yield losses of up to 40% in the first harvesting year have been reported in Malaysia on oil palm (Fee et al. 1999).

Natural enemies

Insects

Only one parasitoid has been reported on the main species of ***Oryctes.*** It belongs to the genus *Scolia* (*Vespoidea Scoliidae*), which includes numerous species: *S. ruficornis* F. in Zanzibar, successfully introduced in Samoa (Simmonds 1961), *S. procer* Ill. in Malaysia, *S. oryctophaga* Coq. in East Africa, etc.
Numerous species of Coleoptera can have a predatory activity on eggs and young larvae. Hoyt (internal report) mentions 14 species, including some *Carabidae*: *Neochryopus savagei* Hope (West Africa), *Scarites madagascariensis*, *Mecodema spinifer* (New Zealand), but also some *Histeridae* (*Pachylister chineensis* Quensel), *Staphylinidae* (*Leptochirus* sp.), etc.
Lastly, we would mention the predators *Platymerus rhadamanthus* Gerst (*P. laevicollis*) (*Heteroptera: Reduviidae*) in East Africa, *P.*

biguttata Stal. and *Psyttala dudgeonia* Dist, these two *Reduviidae* being reported in Nigeria (Jerath 1965).
Most of these species have undergone introductions, notably in the Pacific (Samoa, Fiji) (O'Connor 1961). In view of their non-specificity, these species could have led to ecological imbalances, and are usually not adapted to their new sites. In the best of cases, they have not had any noticeable impact on ***Oryctes*** populations.

Nematodes

The bodies of ***Oryctes*** adults are inhabited by several species of nematodes (Poinar 1970), which do not seem to have any notable impact on the behaviour of the insects. Biological control has been attempted, without success, using *Steinernema feltiae* (Agridos) (*Neoaplectana carpocapsae* Weiser) (Zelazny 1985).

Diseases

Fungi

In larva sites, it is not rare to find ***Oryctes*** larvae or adults affected by the fungus *Metarrhizium anisophiae* (Metsch.), but this fungus remains without any incidence on populations under normal conditions. Nevertheless, severe infestation has been observed locally, such as in India during the monsoon season (Nirula *et al.* 1955, Radha *et al.* 1956). Under a biological control programme, a great deal of research has been carried out, both in the laboratory (Ferron *et al.* 1975 for example) where the fungus proved to be highly aggressive, and under natural conditions (Latch and Falloon 1976, Concibido and Abad 1986, Abad *et al.* 1986).
A mathematical model has been developed for biological control of ***O. rhinoceros*** (Hochberg and Waage, 1991) and has shown that the fungus helps in substantially reducing the juvenile instars of the insect, but does not have any effect on the economic importance of healthy, feeding adult populations. The combined effect of *Metarrhizium* and *Baculovirus* (see below) is greater in wet periods (Villacarlos and Betonio 1990).
Very locally, the fungus *Cordyceps* can have an impact on *Oryctes* populations (Bedford 1968), or has occurred occasionally in a rearing situation (Hurpin 1966).

Virus

Baculovirus oryctes (Hüger) was detected in Malaysia at the beginning of the 60s (Hüger 1966). That discovery led on to a great deal of research, which has been abundantly referred to in the literature (Bedford 1980) and has continued to increase over the last 20 years. The virus was introduced into Africa and proved in Côte d'Ivoire to be substantially more virulent on ***O. boas*** than on ***O. monoceros***, which itself is a much less susceptible species than ***O. rhinoceros*** (Julia and Mariau 1976b). However virus releases in Tanzania showed that ***O. monoceros*** damage dropped by half a year after the release, whilst attacks remained constant in regions where no releases were carried out (Purrini 1989). However, ten years later Seguni *et al.* (1999) stated that there was no significant reduction in damage after virus establishment.

The virus has been released on a very large number of Pacific islands, usually with success. After such a release on Wallis and Futuna, an 82% reduction in attacks was seen, with more substantial reductions in dense coconut plantings (Hammes and Monsarrat 1974). The same applied on Tonga, where there was a considerable reduction in populations and the populations were still at a low level seven years later (Young and Longworth 1981). In Papua New Guinea, the introduction was a success but the impact on populations is unknown (Gorick 1980). *Baculovirus* preparations have been developed so that farmers can easily use the virus (Zelazny *et al.* 1987). In the Maldives and on most of the islands where the virus has been released, a highly significant reduction in ***Oryctes*** damage has been seen (Zelazny *et al.* 1990).

A study of ***O. rhinoceros*** population dynamics in the Philippines showed that fluctuations in its populations were often inversely proportional to viral disease incidence (Zelazny and Alfiler 1991), which only has a slight effect on the survival of the insect's larvae, but considerably reduces the life span of the adults (Zelazny *et al.* 1992). However, we saw in Indonesia that it is very rare to find ***O. rhinoceros*** larvae affected by the virus diseases in larva sites. In eastern Java and southern Sulawesi, Zelazny *et al.* (1989) also observed that the disease existed but cases were very rare and it was thought that the host might be resistant to the virus in those zones; such resistance has been detected in the laboratory in the Philippines. Moreover, it was found that ***O. rhinoceros*** adults from Indonesia (Sumatra) reared in France died rapidly and an analysis of samples revealed the presence of *Baculovirus* (Morin, personal communication), though that does not prevent the occurrence of spectacular outbreaks. Lastly, *Baculovirus* was found to be inactive after spending 8 days in organic matter (Mohan 1991).

Other organisms

A gregarine, *Ophryocystis oryctesi* Purrini and Kohring (1987), has been discovered, along with a Microsporidium, *Pleistophora tanzaniae* Purrini and Kohring (1986), in ***O. monoceros*** but these organisms do not appear to have any adverse effect on host insect development. Other diseases, notably bacterial, have been studied (Surany 1960).

Attractants

Kairomones, along with sexual and aggregation pheromones play a very important role in the chemical ecology of the *Scarabaeidae* (Leal 1998). Among 93 chemicals tested, ethyl dihydrochrysanthemumate has proved to be the best attractant (Barber *et al.* 1971, Van der Meer and Mc Govern 1983). The efficacy of that substance has been confirmed on both ***O. rhinoceros*** (Maddison *et al.* 1973) and ***O. monoceros***, though the size of catches is highly variable depending on environmental conditions (Julia and Mariau 1976c).
An aggregation pheromone (ethyl 4-methyloctanoate) was discovered first in ***O. monoceros*** (Gries *et al.* 1994) then in ***O. rhinoceros*** (Hallet *et al.* 1995, Morin *et al.* 1996). In the latter species, the pheromone has a much greater attractant effect than ethyl chrysanthemate and mainly catches females. The odours are picked up by the antennae, whose structure has been studied (Renou *et al.* 1998). The degree of attraction can be multiplied three-fold by adding to the trap a piece of empty bunch recently ejected from the mill (Desmier de Chenon, personal communication) or rotting coconut wood (Alfiler 1999).
Oryctes have a stridulation apparatus, whose role is not precisely understood (Mini 1997).

Control methods

Oryctes control is primarily preventive, by destroying potential larva sites, or covering them with a cover crop. This technique has proved its merits for both ***O. rhinoceros*** (Wood 1969) and ***O. monoceros*** (Mariau and Calvez 1973, Julia and Mariau 1976a). Curative chemical control can be carried out by applying an insecticide to the axils of young fronds. Applications of Carbofuran have been found to reduce ***O. rhinoceros*** attacks (Toh et Brown 1978) and excellent results have been obtained with Carbosulfan (Desmier de Chenon, personal communication). Although using *Baculovirus oryctes* has led to spectacular results in some cases, notably on islands, in many cases elsewhere its efficacy seems to have remained low. Lastly, pheromone traps can be used to catch a large number of ***Oryctes*** adults (Fee 1997), leading to population decreases (Desmier de Chenon *et al.* 1998), but the percentage of insects captured is unknown. Research under way should lead to an improvement in trap efficacy, with the addition of a further attractant.

Scapanes australis Boisduval

Description

The adult is 6 cm long. Its elytra are shiny dark brown and the thorax almost black. Males have three powerful horns: one cephalic pointing backwards, and two lateral thoracic pointing forwards.

The larvae are typical white grubs and can be distinguished from the larvae of other *Scarabaeidae* frequenting the same sites (Bedford 1974, Beaudoin-Ollivier *et al.* 1999).

There are four sub-species that are clearly separated geographically (Endrödi 1957, Beaudoin-Ollivier *et al.* 1999):

- ***S. australis australis*** (Boisduval) on the northern coast of the Papua New Guinea mainland;
- ***S. australis brevicornis*** (Sternberg) in the Southeast of the mainland (Milen bay);
- ***S. australis salomonensis*** (Sternberg) on the island of Bougainville and in the Solomon Islands;
- ***S. australis grossepunctatus*** (Sternberg) in New Britain and New Ireland (PNG).

Biology

The average duration of the development cycle is 1 year (***S.a. grossepunctatus***), including 290 days for the larva cycle and 45 days for pupation (Bedford 1976). Coconut stumps and old *Gliricidia sepium* stumps are ideal larva sites (Beaudoin-Ollivier, personal communication). Males are mostly found in galleries mined in young coconut tissues. Mating takes place at the same site (Beaudoin-Ollivier *et al.* 1991), unlike with ***Oryctes*** where it takes place in larva sites. In a feeding gallery, the female is almost always accompanied by a male and the sex-ratio, at that level, is highly in favour of males (Bedford, 1979). It is possible to attract ***Scapanes*** adults into a trap containing males, suggesting the existence of a pheromone (Kakul *et al.* 1999, Morin *et al.* 1999, Prior *et al.* 2000), which has now been detected and identified (Rochat *et al.* 1999).

Natural enemies

Fungus strains, *Metarrhizium anisopliae* and *Beauveria bassiana*, have been isolated and are active under laboratory conditions (Beaudoin-Ollivier *et al.* 1999, Prior and Arura 1985), but these pathogens are only rarely found under natural conditions. The first larval instars are susceptible to *Baculovirus oryctes* (Bedford 1973), but it has not been possible to use the virus as a control method. The nematode, *Rhabditis adenobia* (Poinar), isolated from the bursa copulatrix (Poinar 1971) does not seem to affect adult behaviour. *Steinernema glaseri* has been effectively tested in the laboratory on 3rd instar larvae (Beaudoin-Ollivier, personal communication).

Strategus

The genus contains no fewer than 30 species, at least 6 of which can attack *Elaeis* and *Cocos* (Lepesme 1947).

Strategus aloeus L.

This species has been observed in numerous Latin American countries. The adult is 6 cm long. The male has 3 powerful but quite short horns, one cephalic and two thoracic. The development cycle lasts 11 months, including 8 for the larva cycle (Hurpin and Mariau 1966, Genty *et al.* 1978). Attacks are similar to those by ***Oryctes pyrrhus*** in Madagascar. Adults penetrate young palms after mining a gallery in the soil near the plant. More than 5 adults may be found on the same palm (Genty *et al.* 1978); this behaviour is linked to an aggregation phenomenon resulting from male pheromone emission (Rochat *et al.* 1999), meaning that trapping can be developed as a control method. *Baculovirus oryctes* can multiply in the body of the adult (Lomer 1987). However, neither the LD 50 nor the possibility of using this virus as a biological control method are known.

Strategus quadrifoveatus (Palisot de Beauvois)

The adult has horns that are more developed than those of ***S. aloeus*** and the larva cycle is very long (14 months). Substantial damage has been observed in Puerto Rico, particularly after the passage of a cyclone which, after several months, led to an increase in larva sites for the insect (Plank 1948).

Several other species of ***Strategus*** have been reported on *Cocos nucifera,* such as ***S. anachoreta*** Burm. in Trinidad and Tobago (Bondar 1940). Also worth mentioning are ***S. simson*** L. in Jamaica and the Dominican Republic, ***S. barbigerus*** Chap. in Puerto Rico (Lepesme 1947) and ***S. validus*** Fabr. on coconut in Brazil (Quindere *et al.* 1977).

Augosoma centaurus Fabricius (Vénard-Combes and Mariau 1983)

Species found in western and central Africa. The adult, which can reach 8.5 cm, is a shiny mahogany brown colour. The males are larger in size and have two long horns: one cephalic pointing backwards, the other thoracic pointing forwards.

The development cycle lasts 2 years. Larva sites are found in very humid places, the favourite environment being rotting raphia stems. Damage to oil palm and coconut, which is more attractive, is similar to that caused by ***Oryctes.*** Mass flights are seen in wet periods, with congregations of adults, which suggests the existence of a pheromone. Adults are greatly attracted by light, which may prove to be an effective control method.

Xylotrupes

The genus comprises 6 species (Lepesme 1947). Whilst the larvae can develop in rotten palm wood, the adults do not often attack these hosts and are therefore minor pests. We would merely mention ***X. lorquini*** Schauf and ***X. gideon*** L. These two species come in various forms depending on the site. The former is the one most studied. Its development cycle lasts 8.5 months and the adult feeds occiasionally off *Cocos* and *Elaeis* inflorescences (Bedford 1975).

Other species of *Dynastinae*

Papuana woodlarkiana Montr. (*P. armicollis* Frm, *P. semistriata*, *Pimelopus tenuestriatus* Aulm., *Oryctes preussi* Aulm.) (Lepesme, 1947) attacks coconut roots in East New Britain (Bedford, 1980). Many other species are reported on *Cocos,* but their biology is often unknown or they do not cause any notable damage (Lepesme, 1947). We would, however, mention ***Chalcosoma atlas*** L., which attacks the root bulb in Asia (Gabriel 1976), ***Oryctoderus latitarsis*** Burm in Melanesia, ***Temnorrhynchus coronatus,*** a stem borer in Mozambique (Saraiva *in* Lepesme, 1947), ***Trichogornphus semmelinki*** Ritz. in the Solomon Islands (Froggatt 1936 *in* Lepesme 1947), ***T. vicinus*** Dechambre, ***Phyleurus didymus*** F., ***Megasoma anubis*** Chevz. (*Dynastes hector*), which attacks the root bulb, and ***P. castaneus*** Hald. In Brazil (Bondar 1940). Lastly, ***Papuana loevipennis*** Arr. attacks young coconut palms in the Solomon Islands, and ***P. hübneri*** Frm. in the Bismarck Islands (Simmonds 1938).

SCARABAEIDAE REFERENCES

Abad R.G., Aterrado N.C. San Juan, Concibido E.C. (1986). Utilization of the green muscaridine fungus, *Metarrhizium anisophiae* (Metsch) Sorokin against the rhinoceros beetle, *Oryctes rhinoceros* L. (Coleoptera: Scarabaeidae: Dynastinae). II – Laboratory pathogenicity trials and field valuation. *Annual report. Agricultural research Philippine Coconut Authority.*

Abraham V.A., Mohandas N. (1988). Biology of coconut white grub *Leucopholis coneophora* Burm. (Melolonthinae: Scarabaeidae: Coleoptera). *Journal of Plantation Crops* 16 (1): 38-44.

Alfiler A.R.R. (1999). Increased attraction of *Oryctes rhinoceros* aggregation pheromone, Ethyl 4– Méthyloctanoate, with coconut wood. *CORD* 15 (2): 106-114.

Alibert H. (1938). Etude sur les insectes parasites du palmier à huile au Dahomey. *Rev. Bot. Appl.* 18 (207): 753-755.

Barber I.A., McGovern T.P., Beroza M., Hoyt C.P., Walker A. (1971). Attractant for the Coconut Rhinoceros Beetle. *J. Econ. Ent.* 64 (5): 1041-1044.

Beaudoin-Ollivier L., Morin J.P., Prior R.N.B., Kakul T., Ollivier J., Rochat D., Mariau D. (1999). Le complexe *Scapanes–Rhynchophorus*, principal problème entomologique du cocotier en Papouasie-Nouvelle-Guinée. *Plantations, Recherche, Développement* 6 (1): 46-55.

Beaudoin-Ollivier L., Prior R.N.B., Laup S. (1998). A field key to identify some rhinoceros and other beetle larvae breeding in coconut palm habitats in Papua New Guinea. *PNG J. Agric. Forest. and Fish.* 41 (2): 1-15.

Bedford G.O. (1968). Observations of the ecology of *Oryctes* (Coleoptera: Scarabaeidae: Dynastinae) in Madagascar. *Bull. Ent. Res.* 58 (1): 83-105.

Bedford G.O. (1973). Experiments with the virus *Rhabdionvirus oryctes* against the Coconut Palm Rhinoceros Beetles *Oryctes rhinoceros* and *Scapanes australis grossepunctatus* in New Guinea. *Journal of Invertebrate Pathology* 22 (1): 70-74.

Bedford G.O. (1974). Descriptions of the larvae of some rhinoceros beetles associated with coconut palm in New Guinea. *Bull. Ent. Res.* 63: 445-472.

Bedford G.O. (1975). Observations on the biology of *Xylotrupes gideon* in Melanesia. *J. Aust. Entomol. Soc.* 14: 213-216.

Bedford G.O. (1976). Observations on the biology and ecology of *Oryctes rhinoceros* and *Scapanes australis* (Coleoptera: Scarabaeidae: Dynastinae) pests of coconut palm in Melanesia. *J. Aust. Ent. Soc.* 15 (3): 241-251.

Bedford G.O. (1980). Biology, Ecology and Control of Palm Rhinoceros beetles. *Ann. Rev. Entomol.* 25: 309-339.

Bondar G. (1940). Insetos nocivos e molestias do coqueiro (*Cocos nucifera*) no Brasil. *Tipographia naval, Bahia,* 160 p.

Concibido E., Abad R.G. (1986). Utilization of the green muscaridine fungus, *Metarrhizium anisophiae* (Metsch) Sorokin against the rhinoceros beetle, *Oryctes rhinoceros* L. (Coleoptera: Scarabaeidae: Dynastinae). I – Fungus mas production. *Annual report. Agricultural research*, Philippine Coconut Authority.

Desmier de Chenon R., Ginting C.U., Asmady, Sudharto P.S., Sipayung A. (1998). Importance of the Rhinoceros beetle in mature oil palm plantations for effective biocontrol. *International Oil Palm Conf. Bali, Indonesia,* Sept. 1998, 14 p.

Endrödi S. (1957). Rassenbrers von Scapanes australis Boisd. (Coleopt. Dynastinae). *Bonn. Zool. Beitr.* 8: 71-74.

FAO (1964). List of insect pests of coconut palm. Technical working party on coconut production. *Colombo working paper 64/11*, 17 p.

Fee C.G. (1997). The Bioefficacy of the Aggregation Pheromone in Mass Trapping of Rhinoceros Beetles (*Oryctes rhinoceros* L.) in Malaysia. *Planter* 73 (852): 119-127.

Fee C.G., Siang C.S., Ramalingman B. (1999). Effect of *Oryctes rhinoceros* dammage on young plants during establishment on the early yield of oil palm. Porim International palm oil congress 1-6 feb. 1999. Kuala Lumpur, Malaysia.

Ferron P., Robert P.H., Deotte A. (1975). Susceptibility of *Oryctes rhinoceros* adults to *Metarrhizium anisophiae. J. Invertebr. Pathol.* 25: 313-319.

Frappa C. (1933). Sur *Platygenia barbata* Mac-leay, insecte nuisible au palmier en A.O.F. *L'Agronomie coloniale* 22 (182): 41-44.

Gabriel P.P. (1976). Pests of coconut in the Philippines. *Philippine J. of Coconut Studies* 1 (1): 15-26.

Genty P., Desmier de Chenon R., Morin J.P. (1978). Les ravageurs du palmier à huile en Amérique latine. *Oléagineux* 33 (7): 326-419.

Ghesquière J. (1935). Rapport préliminaire sur l'état sanitaire de quelques palmeraies de la province de Coquilhatville. *Publ. INEAC Bruxelles* n° 22.

Gorick B.D. (1980). Release and establishment of the baculovirus disease of *Oryctes rhinoceros* (L.) (Coleoptera: Scarabaeidae) in Papua New Guinea. *Bull. Ent. Res.* 70: 445-453.

Gressit J.L. (1953). The coconut rhinoceros beetle (*Oryctes rhinoceros*) with particular reference to the Palau islands. *Museum Honolulu,* 157 p.

Gries G., Gries R., Perez A.L., Oehlschlager L.M., Gonzales L.M., Pierce H.D. Jr, Zebeyou M., Kouame B. (1994). Aggregation Pheromone of the African Rhinoceros Beetle, *Oryctes monoceros* (Olivier) (Coleoptera: Scarabaeidae). *Z. Naturforsch.* 49c: 363-366.

Hallet R.H., Perez A.L., Gries G., Gries R., Pierce H.D. Jr, Yue J., Oehlschlager A.C., Gonzales L.M., Borden J.H. (1995). Aggregation pheromone of the coconut rhinoceros beetles, *Oryctes rhinoceros* (L.) (Coleoptera: Scarabeidae). *J. Chem. Ecol.* 21 (10): 1549-1570.

Hammes C., Monsarrat P. (1974). Recherches sur *Oryctes rhinoceros* L. *Cah. Orstom, sér. Biol.* n° 22: 43-111.

Hochberg M.E., Waage J.K. (1991). A model for the biological control of *Oryctes rhinoceros* (Coleoptera: Scarabeidae) by means of pathogens. *Journal of Applied Ecology* 28 (2): 514-531.

Hüger A.M. (1966). A virus disease of the Indian rhinoceros beetle *Oryctes rhinoceros* caused by a new type of insect virus *Rhabdionvirus oryctes* gen. n., sp. n. *J. Invertebr. Pathol.* 8: 38-51.

Hurpin B. (1966). Résultats et perspectives de la lutte biologique contre les *Oryctes. Oléagineux* 21 (6): 1-6.

Hurpin B., Mariau D. (1966). Contribution à la lutte contre les *Oryctes* nuisibles aux palmiers. *C. R. Académie Agriculture France* du 26/1/1966: 178-186.

Jerath M.L. (1965). A list of insects found on palms in Nigeria and their known parasites and predators. *Journal of Nifor* (1965).

Julia J.F., Mariau D. (1976a). Recherches sur l'*Oryctes monoceros* Ol. en Côte d'Ivoire. I – Lutte biologique. Le rôle de la plante de couverture. *Oléagineux* 31 (2): 63-68.

Julia J.F., Mariau D. (1976b).). Recherches sur l'*Oryctes monoceros* Ol. en Côte d'Ivoire.– II Essai de lutte biologique avec le virus *Rhabdionvirus oryctes. Oléagineux* 31 (3): 113-117.

Julia J.F., Mariau D. (1976c).). Recherches sur l'*Oryctes monoceros* Ol. en Côte d'Ivoire.– III Piégeage olfactif à l'aide de chrysanthémate d'éthyle. *Oléagineux* 31 (6): 263-272.

Kakul T., Laup S., Stathers T., Beaudoin-Ollivier L., Morin J.P., Rochat D. (1999). Trapping and description of the mating activity of *Scapanes australis* using live male as bait in Papua New Guinea. *CORD,* special ed., XV (2): 131-138.

Latch G.C.M., Falloon R.E. (1976). Studies on the use of *Metarrhizium anisophiae* to control *Oryctes rhinoceros. Entomophaga* 21: 39-48.

Leal W.S. (1998). Chemical ecology of phytophagous scarab beetles. *Ann. Rev. Entomol.* 43: 39-61.

Lepesme P. (1947). Les insectes des palmiers. Paul Lechevalier Ed. Paris, 903 p.

Lever R.J.A.W. (1969). Les ravageurs du cocotier. *Etudes agricoles de la FAO* n° 77, 196 p.

Lomer C.J. (1987). Infection of *Strategus aloeus* (L.) (Coleoptera: Scarabeidae) and other Dynastinae with *Baculovirus oryctes. Bull. Ent. Res.* 77 (1): 45-51.

Maddison P.A., Beroza M., McGovern T.P. (1973). Ethyl Chrysanthemate as an Attactant for the Coconut Rhinoceros beetle. *J. Econ. Ent.* 66: 591-592.

Mariau D. (1967). Les fluctuations de populations d'*Oryctes* en Côte d'Ivoire. *Oléagineux* 22 (7): 451-454.

Mariau D. (1968). Biologie du comportement alimentaire de l'*Oryctes. Oléagineux* 23 (6): 377-380.

Mariau D. (1970). Etude des déplacements d'*Oryctes monoceros* (Coleoptera, Scarabeidae) à l'aide de l'iridium 192. *Oléagineux* 25 (7): 389-391.

Mariau D., Calvez C. (1973). Méthodes de lutte contre l'*Oryctes* en replantation de palmier à huile. *Oléagineux* 28 (5): 215-218.

Mariau D., Desmier de Chenon R., Julia J.F. (1981). Les ravageurs du palmier à huile et du cocotier en Afrique de l'Ouest. *Oléagineux* 36 (4): 168-228.

Mariau D., Desmier de Chenon R., Sudharto P.S. (1991). Les ravageurs du palmier à huile et leurs ennemis en Asie du Sud-Est. *Oléagineux* 46 (11): 400-476.

Mini A. (1997). Scanning Electron Microscopic Studies of the Stridulatory Apparatus of the Coconut Rhinoceros Beetle *Oryctes rhinoceros* L. (Coleoptera: Scarabeidae). *Entomon.* 22 (1): 7-13.

Mohan K.S. (1991). Persistance of *Oryctes* Baculovirus in organic matter. *J. Biol. Control* 5 (1): 28-31.

Monty J. (1978). The coconut palm rhinoceros beetle *Oryctes rhinoceros* L. in Mauritius, and its control. *Rev. Agr. Sucr. Ile Maurice* 57: 60-76.

Morin J.P., Kakul L., Beaudoin-Ollivier L., Rochat D., Prior R.N.B. (1999). Importance of biological knowledge for trapping. The case of *Scapanes australis* (Dynastidae), a coconut pest in Papua New Guinea. ANPP-5th International Conference on agricultural pests, 7-8-9 December 1999.

Morin J.P., Rochat D., Malosse C., Lettere M., Desmier de Chenon R., Wibwo H., Descoins C. (1996). Le 4-méthyloctanoate d'éthyle, composant principal de la phéromone mâle de *Oryctes rhinoceros* (L.) (Coleoptera: Dynastinae). *C. R. Acad. Sci. Paris* 319: 595-602.

Nirula K.K., Radha K., Menon K.P.V. (1955). The green muscaridine disease of *Oryctes rhinoceros* L. I – Symptomatology, epizootiology and economic importance. *Indian Coconut J.* 9: 3-10.

Nirula K.K. *et al.* (1952). A new pest of coconut palms in India (*Leucopholis coneophora*). *Indian Coconut J.* 4 (4): 225-234.

O'Connor B.A. (1961). Introduction dans le Pacifique de prédateurs et de parasites de l'*Oryctes. Bull. Pacifique Sud* 11 (1): 37.

Pillai R.M. (1921). Short notes on the insect pests of crops in Travancore. *Travancore Dept. Agricul.* Trivandrum, 53 p.

Plank H.K. (1948). Life history, habits, and control of the coconut rhinoceros beetle in Puerto Rico. *Bull. Fed. Exp. Stn. Mayaguez, Puerto Rico* n° 45, 35 p.

Poinar G.O. (1970). Les nématodes, parasites de l'*Oryctes rhinoceros*. *Bulletin du Pacifique Sud* 20 (2): 41-42, repris par *Oléagineux* 29 (7): 465-467.

Poinar G.O. (1971). *Rhabditis adenobia* in the colleterial glands of *Scapanes*. *Proc. Helminthol. Soc. Wash.* 38 (1): 99-108.

Prior C., Arura M. (1985). The infectivity of *Metarrhizium anisopliae* to two insect pests of coconuts. *Journal of Invertebrate Pathology* 45 (2):187-194.

Prior R., Morin J.P., Rochat D., Beaudoin-Ollivier L., Stathers T., Kakul T., Empuba S., Nanguai R. (2000). New aspects of the biology of the Melanesian Rhinoceros Beetle *Scapanes australis* (Col., Dynastidae) and evidence for field attraction to males. *J. Appl. Ent.* 124: 41-50.

Purrini K. (1989). *Baculovirus oryctes* Release into *Oryctes monoceros* Population in Tanzania, with Special Reference to the Interaction of Virus Isolates Used in Our Laboratory Infection Experiments. *Journal of Invertebrate Pathology* 53 (3): 285-300.

Purrini K., Kohring G.W. (1986). Light and Electron Microscopic Studies on a New Microsporidian, *Pleistophora tanzaniae* n. sp. (Microsporida: Microspora) Parasiting the *Oryctes monoceros* Oliv. (Scarabeidae, Coleoptera). *Arch. Protistenkd.* 131: 281-286.

Purrini K., Kohring G.W. (1987). *Ophryocystis oryctesi* n. sp. (Ophryocystidae,, Neogregarinida), a New Gregarine Parasitizing the Rhinoceros Beetle *Oryctes monoceros* Oliv. (Scarabeidae, Coleoptera). *Zool. Anz.* 218 (314S): 237-245.

Quindere W.W.A., Pereira L., Da Silva Carneiro J. (1977). Caracterizaçao da larva do besouro do coquiero; *Strategus validus* Fabr. 1775 (Coleoptera) *Fitossanidade*, Brasil 2 (1): 20-21.

Radha K., Nirula K.K., Menon K.P.V. (1956). The green muscaridine disease of *Oryctes rhinoceros* L. II The causal organism. *Indian Coconut J.* 9: 83-89.

Renou M., Tauban D., Morin J.P. (1998). Structure and function of antennal pore plate sensilla of *Oryctes rhinoceros* (L.) (Coleoptera: Dynastinae). *Int. J. Insect Morphol. & Embryol.* 27 (3): 227-233.

Rochat D., Ramirez-Lucas P., Malosse C., Aldana R., Kakul T., Morin J.P. (1999). Role of solid phase micro-extraction in the identification of highly volatile pheromones of two Rhinoceros Beetle *Scapanes australis* and *Strategus aloeus* (Coloeptera: Scarabeidae: Dynastinae). *J. Chrom. A.* 885: 433-444.

Seguni Z., Lohr B., Mwaiko W. (1999). Introducing *Baculovirus oryctes* Huger into *Oryctes monoceros* Oliv. populations in Tanzania. *J. Appl. Ent.* 123: 427-431.

Simmonds H.W. (1938). Coconut pests and diseases in Melanesia and southern Polynesia. *Depart. of Agric.Fiji. Bull.* n° 20.

Simmonds H.W. (1961). A short history of *Scolia ruficornis*, parasite of *Oryctes* spp. *Agricult. J. Fiji* 31: 36-38.

Surany P. (1960). Diseases and biological control in Rhinoceros beetles. *South Pacific Commission Technical Paper* n° 128 Noumea, 62 p.

Toh P.Y., Brown T.P. (1978). Evaluation of carbofuran as a chemical prophylactic control measure for *Oryctes rhinoceros* in young oil palms. *Planter* 54 (3): 3-11.

Van der Meer R.K., Mc Govern T.P. (1983). Structure-activity correlations for derivative of siglure: attracttants for *Oryctes rhinoceros* L. (Coleoptera: Scarabeidae). *J. Econ. Entomol.* 76: 723-727.

Venard-Combes P., Mariau D. (1983). *Augosoma centaurus* Fabricius (Coleoptera: Scarabeidae) important ravageur du cocotier en Afrique. Descriptions, biologie, méthode de lutte. *Oléagineux* 38 (12): 651-657.

Villacarlos L.T., Betonio P.A. (1990). Incidence of Baculovirus and green muscaridine fungus in the larval population of rhinoceros beetle in Baybay, Leyte, Philippines (a research note). *PJCS* 15 (1): 21-23.

Vinson J. (1963). Sur l'apparition à Maurice d'un scarabée nuisible, *Oryctes rhinoceros* Linné. *Rev. Agric. Sucr. Ile Maurice* 42: 173-180.

Wood B.J., Kwang Yew Ng (1969). The cockchafer, *Psilopholis vestita,* a new pest of oil palm in West Malaysia. *Planter,* 45, 524: 577-586.

Wood B.J. (1969). Studies on the effect of ground vegetation on infestations of *Oryctes rhinoceros* (L.) in young oil palm replantings in Malaysia. *Bull. Ent. Res.* 59: 85-96.

Young E.C., Longworth J.F. (1981). The Epizootiology of the Baculovirus of the Coconut Palm Rhinoceros beetle (*Oryctes rhinoceros*) in Tonga. *Journal of Invertebrate Pathology* 38: 362-369.

Zelazny B. (1985). Susceptibility of two coconut pests, *Oryctes rhinoceros* (Col: Scarabeidae) and *Tirathaba rufivena* (Lep: Pyralidae) to the entomoparasitic nematode *Steinernema feltiae* (= *Neoaplectana carpolapsae*). *Entomophaga* 30 (2): 121-124.

Zelazny B., Alfiler A.R. (1991). Ecology of Baculovirus. Infected and Healthy Adults of *Oryctes rhinoceros* (Coleoptera: Scarabeidae) on coconut palms in the Philippines. *Ecological Entomology* 16 (2): 253-259.

Zelazny B., Alfiler A.R., Crawford A.M. (1987). Preparation of a baculovirus inoculum for use by coconut farmers to control rhinoceros beetle (*Oryctes rhinoceros*). *FAO Plant Prot. Bull.* 35 (2): 36-42.

Zelazny B., Alfiler A.R., Lolong A. (1989). Possibility of resistance to a baculovirus in populations of the coconut rhinoceros beetle (*Oryctes rhinoceros*). *FAO Plant Prot. Bull.* 37 (2): 77-82.

Zelazny B., Lolong A., Crawford A.M. (1990). Introduction and field comparison of Baculovirus strains against *Oryctes rhinoceros* (Coleoptera: Scarabeidae) in the Maldives. *Environ. Entomol.* 19 (4): 1115-1121.

Zelazny B., Lolong A., Pattang B. (1992). *Oryctes rhinoceros* (Coleoptera: Scarabeidae) Populations Suppressed by a Baculovirus. *Journal of Invertebrate Pathology* 59: 61-68.

CHRYSOMELIDAE

The family of the *Chrysomelidae* comprises some 37,000 species, with a different number of subfamilies depending on the authors. We shall adopt Jolivet's classification (1997) in twenty current subfamilies. It is that of the *Hispinae* that is most represented among the species living off *Elaeis guineensis* and especially *Cocos nucifera*. There are also more than fifteen species, including some pests that can be economically important, in the subfamilies *Cassidinae, Eumolpinae* and *Galerucinae* (Mariau 1999).
The *Chrysomelidae* include numerous major agricultural pests in both temperate and tropical countries, of which the most famous is the Colorado Beetle, *Leptinotarsa decemlineata* Say, whilst *Coelaenomenodera lameensis* Berti and Mariau has caused severe defoliation over wide areas of oil palm in western Africa.

CASSIDINAE

Worldwide, this subfamily includes no fewer than 159 genera and some 3,000 species. Fewer than ten of them have been observed on *Cocos* and *Elaeis*, mostly in South America and Africa.

The genus ***Himatidium*** includes several species reported in Brazil by Bondar (1940), the most common being ***H. neivai*** Bondar. The larvae and adults, both of which have a highly flattened appearance, scrape away the leaflet surface of young unopened fronds, and the surface of palm nuts and fruits, on which damage leaves very typical pouches. However, such attacks are not serious. On coconut, Bondar also mentions: ***H. bahianum*** Spaeth and ***H. bondari*** Spaeth.

Delocrania cossyphoides Guérin is a very common species on *E. guineensis* and *C. nucifera* in numerous Latin American countries. The adult has a similar habit to that of ***Himatidium***. Both larvae and adults feed off the underside of leaflets on young palms, where they eat away the epidermis (Genty *et al.* 1978, Zenner 1968), but attacks usually remain limited, probably due to high parasitism, on eggs by a *Hymenoptera Trichogrammatidae*, and on larvae and pupae, parasitized by the Hymenoptera *Eulophidae* and *Ichneumonidae.*

Spaethiella tristis Boheman is dark blue to almost black in colour, and is circular in shape. The elytra are highly sculptured (Genty *et al.* 1978). The larvae are elongated with legs bearing strong hooks, enabling it to cling on to plants. They are covered in a kind of fibrous case; the interwoven fibres are made from its own excreta. In the Ecuadorian Amazon, we observed substantial and sudden outbreaks over extensive areas of oil palm plantings. There does not seem to be much parasitism, with a Hymenoptera *Ichneumonidae* on the final larval instar, a *Eulophidae* of the genus *Horismenus* on pupae and a

Pteromalidae of the genus *Halticopteroides* on larvae (Delvare, personal communication). Another, reddish species, ***S. costipennis*** Boheman, has been seen on *Elaeis* in northern Colombia (Genty *et al.* 1978). The species has been reported on coconut in Brazil, under the genus name ***Haemishaerota*** (Lima 1997).

Calyptocephala brevicornis Boheman has been observed in the State of Amapa, Brazil (Moura 1985). The larvae are parasitized by two species of chalcids.

A few species have been noted in Africa: ***Aspidomorpha obovata*** Klug on *Elaeis* in Sierra Leone (Hargreaves 1937), along with ***A. cincta*** and ***Cassida*** sp. on *Cocos* in Nigeria.

EUMOLPINAE

This is an important subfamily of the *Chrysomelidae* with 500 genera and 3,400 species described, but whose host plants primarily belong to the dicotyledons (over 80%) (Jolivet 1997). The species found on palms are very few in number. On *Cocos*, just a few species have been observed in Brazil: ***Brevicolaspis villosa*** Bryant, ***Metaxionycha hibrida*** Lef., ***M. testacea*** Fabricius. and ***Colaspis*** sp. (Bondar 1940). The adults feed off leaflets and their damage takes the form of lines parallel to the midrib. Damage is minimal in the majority of cases. In Africa, a single species has been reported on *Elaeis*, more particularly in Sierra Leone: ***Syagrus puncticollis*** Bry (Hargreaves 1937).

GALERUCINAE

This subfamily comprises around 5,800 species, which usually attack dicotyledons, like the *Eumolpinae.* In Africa, a few species have been reported on various palms, including three on *Elaeis* and *Cocos*.
They belong to the vast genus ***Monolepta,*** which contains 600 species, including ***M. limbata*** Olivier and ***M. tenuicornis*** Jacoby in Sierra Leone (Hargreaves 1937). The species ***M. marica*** Wilcox (***M. apicicornis*** Jacoby) (Laboissière 1940) is a pest that can have a certain economic impact. This species is found on *Elaeis guineensis* in West Africa, as far as Cameroon (Mariau and Djob Bikoi 1990). The adult, which measures 5 to 8 mm in length, is a mahogany brown colour. The female lays its eggs on the soil surface. The first instar caterpillars penetrate order III and order IV roots, completing their cycle in the primary roots. The root system can be partially destroyed. Two generations are seen per year, with adult emergences clearly marked in time.

This subfamily, which comprises more than 3,000 described species, is primarily tropical. Many species remain undocumented. Of the 110 genera inventoried, almost 80% live off monocotyledons (Jolivet 1997) and numerous species, including some serious pests, live off oil palm and coconut.
The species of the *Hispinae* can be divided into two groups:
- Those that browse the foliage, but live between the leaflets of unopened leaves. They belong to the tribes *Cryptonichini, Alurnini, Cephaloleini* and *Botryonopini.*
- Those whose larvae mine a gallery in the lamina of opened leaves. These belong to the tribes *Coelaenomenoderini, Promecothecini* and *Hispoleptini.*

Leaf-browsing *Hispinae*

The Hispinae revealing this behaviour are mostly found in the zone of origin of the coconut palm, i.e. in Asia and the Pacific Ocean, though there are a few species in South America.

Cryptonichini

Virtually all the species in this tribe live off coconut in Asia and the Pacific zone. Both the larval and adult stages of these insects live between the leaflets of unopended fronds forming the spear. The slower the leaf emission, the greater the damage.
With 7 very similar species, the most common genus is ***Brontispa***: ***B. longissima*** Gestro is economically the most serious (Anon. 1983, Cohic 1961, Stapley 1980).
The adult measures 8 to 12 mm in length. It varies in colour from reddish-brown to black depending on the region. In fact, the species is divided into numerous varieties depending on the islands: **B.l. *longipennis*** Gestro (New Guinea), **B. l. *frogatti*** Sharp (Vanuatu, New Caledonia), **B. l. *javana*** Wise, **B. l. *celebensis*** Gestro, etc (Lepesme 1947). The development cycle varies from 5 to 8 weeks, and fecundity exceeds a hundred eggs. This species is controlled by several chalcid Hymenoptera (Cox 1994), on both eggs: *Hispodiphila brontispae* (Ferrière), *Trichogrammatoidea nana* Zehntner (*Trighogrammatidae*) and *Oencyrtus* sp. (*Encyrtidae*) and on larvae and pupae: *Tetrastichus brontispae* (Ferrière) (*Eulophidae*). These parasitoids were studied with a view to applying biological control methods. They have been used in several introduction trials (Cochereau 1965, 1969, Voegele 1989). Unsuccessful biological control trials have been attempted with the fungus *Metarrhizium anisopliae* (Liu and Lin 1989).
Chemical control is often carried out in nurseries (Brown and Green 1958).

The other species of ***Brontispa*** are: ***B. palauensis*** (Esaki & Chujo) (Muniappan *et al.* 1980), and ***B. chalybeipennis*** Zacher (island of Palau), ***B. mariana*** Spaeth (***B. castaneipennis*** Chujo) in the Mariana Islands (Corbett 1923), ***B. depressa*** Baly and ***B. suriganoa*** Uhm in the Philippines. ***B. limbata*** Waterhouse in Mauritius is the only species living off coconut in the Asia-Pacific zone.

Three other species were long confused with the genus ***Brontispa***: ***Dactylispa angulosa*** (Uhm) and ***Octodonta afinis*** Uhm in the Philippines and ***Calenodispa sarrasini*** (Heller) in New Caledonia (Uhmann 1957).

The *Cryptonichini* include other morphologically distant species (Kalshoven 1957): ***Plesispa reichei*** Chapuis (Maulik 1937), widespread in Indonesia and Malaysia, whose eggs are parasitized by *Ooencyrtus podontiae* Gahan (*Encyrtidae*) and, like ***B. longissima,*** by *Hispodiphila brontispae* (Lever 1969), and the larvae and pupae by *Pediobius imbreus* (Walker) and *Tetrastichus plesispae* Ferrière (*Eulophidae*) (Corbett 1923). Two other species, ***P. papuana*** Gestro and ***P. ruficollis*** Spaeth are limited to New Caledonia.
Callispa sp., whose development cycle lasts 1.5 months, is considered to be a minor coconut pest in Kerala (India) (Nair and Oommen 1965).
Isopedhispa cocotis Maulik exists in New Caledonia (Risbec 1934).
Pistosia testacea Fabricius has been reported in New South Wales.
Callistola tripartita Fairmaire and ***Aulostyrax nuciferae*** Maulik are from the Solomon Islands (Froggatt 1936, Maulik 1937).
Froggatt (1936) mentioned three species of ***Oxycephala*** causing similar damage to ***Brontispa*** in Papua New Guinea: ***O. papuana*** (***P. papuana)***, ***O. cornigera*** Guérin and ***O. wallacei*** (***Callistola tripartita*** Fairmaire).

Several species have been reported outside the Asia-Pacific zone:

Cryptonychus cryptonychellus breviceps Weise (Cameroon and Congo) with a variety ***palmarum*** Maulik, mentioned in Ghana. Unlike all the other species of *Cryptonichini*, the damage is that of a leaf miner (Lepesme 1947).

The genus ***Gestronella*** is Madagascan. Of the six species inventoried, two have been observed on coconut, ***G. centrolineata*** (Fairmaire) and ***G. lugubris*** (Fairmaire) (Fairmaire 1890, Appert 1974, Uhmann 1963). A parasitoid of the genus *Pediobius* is found in the final larval instar and pupal stages. On the eastern coast of Madagascar, ***G. centrolineata*** sometimes causes serious damage in young plantings, whereas ***G. lugubris*** has only been seen on mature coconut palms (Mariau, unpublished).

Alurnini

In this tribe, three species of the genus ***Coraliomela*** have been reported: ***C. brunnea*** Thunberg on coconut in Brazil, which can cause substantial damage on young palms. Three parasitoids have been inventoried on eggs: one belonging to the genus *Tetrastichus* and two to the genus *Closterocerus (Eulophidae)* (Ferreira and Morin 1986). The other two species are also found on *Cocos* in Brazil: ***C. quadrimaculata*** Guérin and ***C. marginata*** Latr. (Bondar 1940).

Alurnus humeralis Rosenberg (Santos 1968, Genty *et al.* 1978)
The adult, which is the largest *Hispinae* known on palms, measures 35 mm in length. It has a black head, red thorax and greenish-yellow elytra with two black dots. The development cycle lasts approximately one year (incubation: 29 to 43 days, 8 larval instars: 224 to 271 days, pupa: 26 to 37 days). Given its large size, the insect can cause substantial damage to young oil palms in Ecuador and Colombia.
Also worth mentioning is the genus ***Platyauchenia*** with ***P. latreillei*** Castelnau on *Cocos* in Brazil (Bondar 1940, Maulik 1940) and ***P. deyrollei*** Guérin also in Brazil (Bondar 1924 *in* Lepesme 1947).

Cephaloleini

The most important species from an economic viewpoint is ***Demotispa*** aff. ***pallida***, which is sometimes confused with ***Himatidium neivai***. The adult measures 5 mm and is reddish brown-in colour. The larvae and adults, which have a highly flattened shape, eat away the outer surface of oil palm fruits (Genty *et al.* 1978). The pupae are parasitized, though rarely, by a *Tetrastichus* and by a chalcid of the genus *Conura* (Genty and Mariau 1973). The larvae are parasitized by a *Eulophidae* and the pupae by *Conura adela* (Burks) (Delvare 1992).

Three species of the genus ***Cephaloleia***, which comprises over 180 species (Uhmann 1957), have been observed on *Elaeis guineensis* in Brazil: ***Cephaloleia elaeidis*** Maulik (Maulik 1937), and more recently ***C. depressa*** Baly, whilst another species, ***C.*** nr. ***vagelineata***, Pic is common in numerous Latin American countries (Sandino 1972).

Lastly, a *Cephaloleini* of the genus ***Stenispa*** has been seen in very large numbers on oil palm in Colombia (Genty, personal communication).

Botryonopini

This tribe only includes one species on coconut, in Java: ***Botryonopa sanguinea*** Guérin (Dammerman 1929).

Leaf-mining *Hispinae*

This group of *Hispinae* includes several serious oil palm and coconut pests in Africa, Asia and South America. The female lays its eggs on the underside of fronds, inserting them into the leaf. The dorsally-ventrally flattened larvae all have the same face; they mine a gallery within the leaf which is clearly visible, forming a ridge on the upperside.

Coelaenomenoderini

The genus ***Coelaenomenodera*** (Fairmaire 1890,Gestro 1908) comprises 39 species, 32 of which are Madagascan (Uhmann 1957). Three of them attack oil palm in Africa: ***C. speciosa*** Gestro in the Congo, ***C. elaeidis*** Maulik, widespread in western and central Africa, reported by Maulik for the first time in 1920, then by Cotterell in 1925, is now synonymous with ***C. minuta*** (Uhmann 1957); lastly, ***C. lameensis*** Berti and Mariau (1999), which was long confused with the previous two.

Coelaenomenodera lameensis

Description
The adult is yellowish-orange; it is 4.5 to 6 mm in length. The fore section of the thorax extends over the head.
The larva measures 6.5 mm in the final instar and is yellow in colour. The head is embedded in two lobes of the thorax.

Behaviour and biology
This is the most serious oil palm pest throughout West Africa, up to and including Cameroon. Outbreaks affecting thousands or even tens of thousands of hectares of *Elaeis guineensis* have been observed in most of the countries in this region (Cachan 1957). The larvae have great difficulty developing on the hybrid derived from the cross between *E. guineensis* and *E. oleifera* (Philippe 1977). It is due to its economic importance that particular attention has been paid to studying this insect.

Development cycle

The cycle lasts around 95 days (incubation: 20 days, 4 larval instars: 44 days, pupa: 11 days, preoviposition period: 20 days) (Morin and Mariau 1970).

Fecundity

Under rearing conditions, the number of eggs laid per female varies from around a hundred to over 500 depending on experimental conditons (Morin and Mariau 1971). Under natural conditions, fecundity is found to greatly depend on climatic conditions, notably the temperature and humidity. It can be virtually zero or exceed 300 eggs per female during a laying cycle (Mariau and Lecoustre 2000). Egg viability is closely linked to climatic factors, particularly temperature and relative humidity (Mariau and Lecoustre 2001). On average, the fecundity of ***C. elaeidis*** is a third of that for ***C. lameensis*** (Mariau 2001).

Parasitoids (Waterston 1925, Cotterell 1925, Mariau and Morin 1971, 1972, 1974, Mariau *et al.* 1978, Morin and Mariau 1974, Mariau and Lecoustre 2001, Mariau 1988).

The parasite complex includes three species on eggs: *Closterocerus leptocerus* (Waterston) (*Eulophidae*), which is the most important, and two species of *Oligosita: O. robusta* and *O. minuta* Viggiani (Viggiani 1980) (*Trichogrammatidae)* and four species, all belonging to the *Eulophidae,* on larvae and pupae: *Pediobius setigerus* Kerrich, which is the most important, is itself parasitized by *Pediobius coffeicola* (Ferrière) and *P. vigintiquinque* Kerrich (1970), *Sympiesis aburiana* (Waterston) and *Pleurotroppopsis podagrica* (Waterston). The fourth species, *Closterocerus africanus* Waterston, is very rare. Lastly, we would mention a *Braconidae* parasitoid on adults, *Perilitus persimilis* De Saeger, which is also very rare. A comparative study of ***C. lameensis*** and ***C. elaeidis*** fecundity and mortality was carried out; it explained the frequent outbreaks of the former and the population stability of the latter (Mariau 2001).

Predators

Ants play a very important role in regulating ***C. lameensis*** populations by destroying a fair number of larva galleries. The main species are*:* *Crematogaster gabonensis* (Emery), *Oecophylla longinoda* (Latreille), *Tetramorium aculeatum* (Mayr.), *Pheidole megacephala* (Fabricius) and *Monomorium* sp. (Mariau *et al.* 1978, Dejean *et al.* 1997, Mariau and Lecoustre 2001).

Entomopathogenic organisms

A fungus belonging to the genus *Tilachlidiopsis,* described by Jover (1950) under the name *Stilbella,* is frequently seen on adults, usually without any major impact on populations (Philippe and Mariau 1996).

Two viruses have been isolated from larvae, one is a densovirus, the other is a little larger (40 nm) (Fédière *et al.* 1986). Insofar as larva mortality follows an epidemic pattern, these viruses seem to play a very important role in regulating populations, though it has not been clearly demonstrated (Mariau and Lecoustre 2001).

Control method
It is by developing a rational chemical control method that populations can be controlled (Mariau and Philippe 1983, Philippe 1990a). The insecticide thiocyclam hydrogen oxalate (Evisect S) gives excellent results (Philippe 1990b) when sprayed against adults. Treatments by stem injection or root uptake of monocrotophos give a very high mortality rate on larvae and adults (Mariau and Philippe 1983).

Coelaenomenodera perrieri

Of all the Madagascan species of ***Coelaenomenodera***, only one is found on coconut, ***C. perrieri*** Fairmaire (Fairmaire 1898, Mariau 1975). Some African parasitoids are found on this species: *Oligosita longiclavata* on eggs, *Sympiesis aburiana, Pleurotroppopsis podagrica, Closterocerus* sp. and another *Eulophidae* of the genus *Chrysonotomyia* Ashmead. Given its wide spectrum of activity on all larval instars, this parasite was introduced into Côte d'Ivoire to ensure more effective control of ***C. lameensis,*** but it was unable to adapt to its new host (Lecoustre *et al.* 1980).

The genus ***Balyana***, which is found almost exclusively in Madagascar, includes one species on coconut (Mariau 1975) ***B. mariaui*** Berti and Desmier de Chenon (1987). This species, which is found mixed with ***C. perrieri,*** has the same parasite complex as the latter, along with *Pediobius balyanae* Bouček (1976).

Promecothecini

This tribe only has one main genus ***Promecotheca***, represented by 24 species in the Asia-Pacific zone, 10 of which have been observed on *Cocos nucifera* (Lepesme 1947, Uhmann 1957, Lever 1969).
The most economically important species is ***P. cumingii*** Baly (Ding Siew Ming 1976; Perera 1982), which is widespread in Malaysia, Indonesia and the Philippines, where there have been outbreaks (Lever 1969). In 1972, the *Hispinae* was introduced into Sri Lanka (Fernando 1972). The parasite complex of this species has been studied, notably by Galego (1986) (table 5). Recently the Hymenoptera *Braconidae, Aneuradesha harle*yi Quicke has been observed as an ectoparasitoid on larvae (Quicke and Polaszek 2000).
The other species existing on coconut are ***P. callosa*** Baly and ***P. varipes*** Baly in Queensland, ***P. coeruleipennis*** Blanchard in Fiji and Tonga (called ***P. reichei*** Baly particularly in the Philippines and ***P. lindingeri*** Aulmann in Samoa). To these can be added ***P. nuciferae*** Maulik in Sulawesia, ***Promecotheca papuana*** Csiki in Papua New Guinea and the Solomon Islands, ***P. opacicollis*** Gestro in Vanuatu, and lastly ***P. soror*** Maulik in the Moluccas. All these species have numerous parasitoids (Kerrich 1974, Cox 1994) (table 1), some of

which are common to several species. *Pediobius parvulus* Ferrière was successfully introduced in Fiji from Java to control ***P. coeruleipennis*** (Taylor 1937), in Srri Lanka against ***P. cumingii*** (Dharmadhikari *et al.* 1977), then in Papua New Guinea (Kalshoven 1981), but did not adapt to ***Coelaenomenodera lameensis*** after its introduction in Côte d'Ivoire (Philippe *et al.* 1979).
To these parasitoids should be added the predators, which include two species of *Pentatomidae* of the genus *Amyotea: A. reciproqua* Wlk. and *A. antigua* Wse. (Kowalski 1917), and the mite *Pyemotes ventricosus* Newport (Lever 1969).

Hispoleptini

This tribe only has one genus, ***Hispoleptis*** Baly (Aslam 1965), with various species that have adapted well to oil palm in South America. Depending in particular on the colour of the elytra and the pronotum, four species have been described (Genty *et al.* 1978, Berti and Desmier de Chenon 1973, Le Verdier and Genty 1988).

Hispoleptis diluta (Guérin-Meneville) described in French Guiana, ***H. subfasciata*** Pic, from the Amazon region, ***Hispoleptis elaeidis*** Aslam, from Ecuador and ***H. ollagnieri*** Berti and Desmier de Chenon from Colombia. Apart from ***H. diluta***, the following parasite complex has been found on the other species: on eggs, an *Oligosita* (*Trichogrammatidae*), two *Eulophidae, Emersonella palmae* Bouček (1977) and *Nesolynx* sp. and, much rarer, an *Encyrtidae*. On larvae, the best known parasitoids are a *Bracon* sp., a *Tetrastichus* sp., both unsuccessfully introduced in Côte d'Ivoire (Philippe *et al.* 1979), and more secondarily two *Ichneumonidae, Cryptinae* and *Pimplinae.* Lastly, the pupae and prepupae are parasitized by *Chalcididae*, of the genus *Conura* sp., *C. desmieri* Delvare, and *C. hispinephaga* Delvare and a *Brachymeria* (Desmier de Chenon 1973, Mariau 1988).

CONCLUSION

On oil palm and coconut, around 70 species of *Chrysomelidae* have been inventoried, classed in four subfamilies. Almost 75% of them belong to the subfamily of the *Hispinae*. It is among these that very serious grade one pests are found, including ***Coelaenomenodera lameensis*** in Africa on oil palm, but also ***Promecotheca cumingi*** and ***Brontispa longissima*** in the Asia-Pacific region on coconut, along with ***Spaethiella tristis, Alurnus humeralis*** and ***Hispoleptis*** spp. on oil palm in South America. Most of the species attack leaves, such as browsers on unopened fronds, or leaf miners in opened fronds, but a few species are found on fruits and roots. The many species of parasitoids, mostly belonging to the order of the Hymenoptera and the of the Hymenoptera and the family of the *Eulophidae,* are not always enough to check outbreaks. That is why parasitoids are introduced, sometimes successfully, primarily in Asia for species of the genera ***Brontispa*** and ***Promecotheca,*** but introductions carried out against ***Coelaenomenodera lameensis*** in West Africa have been unsuccessful.

Table 5. Parasitoids of *Promecotheca* spp.

Promecotheca Species	Parasite stages	Parasitoids	
		Families	Genera and species
P. cumingii	Egg	*Eulophidae*	*Closterocerus promecothecae* (Ferrière)
	Egg	*Aphelinidae*	*Centrodora* sp.
	Larva	*Eulophidae*	*Cirrospilus* sp.
	Larva	*Eulophidae*	*Pediobius anomalus* (Gahan)
	Larva	*Eulophidae*	*Sympiesis javanica* (Ferrière)
	Larva + pupa	*Eulophidae*	*Pleurotroppopsis lankensis* (Kerrich) (in Sri Lanka)
Promecotheca papuana	Egg	*Eupelmidae*	*Anastatus* sp.
	Egg	*Eurytomidae*	*Eurytoma promecothecae* Ferrière
	Egg	*Trichogrammatidae*	*Oligosita utilis* Kowalski
	Larva	*Eulophidae*	*Pediobius parvulus* (Ferrière)
	Larva	*Eulophidae*	*Closterocerus splendens* Kowalski
	Larva	*Eulophidae*	*Apleurotropis lalori* Girault
	Larva	*Eulophidae*	*Derostenus* sp.
	Larva	*Eulophidae*	*Eutedon* sp.
P. opacicollis	Egg	*Trichogrammatidae*	*Oligosita utilis* Kowalski
	Larva	*Eulophidae*	*Closterocerus splendens* Kowalski
	Larva	*Eulophidae*	*Entedon* sp.
P. reichei	Egg	*Trichogrammatidae*	*Oligosita utilis* Kowalski
	Larva	*Elasmidae*	*Elasmus hispidarum*Ferrière
P. coeruleipennis	Egg	*Tricogrammatidae*	*Oligosita utilis* Kowalski
	Larva	*Elasmidae*	*Elasmus hispidarum* Ferrière
	Larva	*Eulophidae*	*Pediobius parvulus* (Ferrière)

CHRYSOMELIDAE REFERENCES

Anonymus (1983). Le Brontispe du cocotier *Brontispa longissima* (Gestro). Commission du Pacifique Sud (Fiche technique), 17 p.

Appert J. (1974). Sur deux coléoptères Hispines du genre *Gestronella* nuisibles au cocotier à Madagascar. *Oléagineux* 29 (12): 559-564.

Aslam N. A. (1965). On *Hispoleptis* Baly (Coleoptera,Hispidae) and *Imatidium* F. (Coleoptera, Cassididae). *Annals and Magazine of Natural History* Ser. 13 Vol. VIII: 687-693.

Berti N., Desmier de Chenon R., (1973). Coléoptères Hispinae d'Amérique du Sud nuisibles aux palmiers. Description d'une nouvelle espèce. *Bulletin de la Société entomologique France* 80: 192-198.

Berti N., Desmier de Chenon R. (1987). Le genre *Balyana* Péringuey Taxonomie et données biologiques (Col. Chrysomelidae Hispinae). *Bulletin Société entomologique France* 92 (3-4): 79-101.

Berti N., Mariau D. (1999). *Coelaenomenodera lameensis* n.sp., ravageur du palmier à huile (Coleoptera: Chrysomelidae) *Nouvelle Revue d' Entomologie* 16 (3): 253-267.

Bondar G. (1940). Insetos nocivos e molestias do coqueiro (*Cocos nucifera*) no Brasil. Bahia, *Tipographia naval*, 160 p.

Bouček Z. (1976). Taxonomic studies on some *Eulophidae* (Hym.) of economic interest, mainly from Africa. *Entomophaga* 21 (4): 401-414.

Bouček Z. (1977). Descriptions of two new species of neotropical Eulophidae (Hymenoptera) of economic interest, with taxonomic notes on related species and genera. *Bull. Ent. Res.* 67: 1-15.

Brown E.S., Green A.H. (1958). The control by insecticides of *Brontispa longissima* (Gestro) (Coleopt., Chrysomelidae-Hispinae) on young coconut palms in the british Solomon islands. *Bull. Ent. Res.* 49: 239-272.

Cachan P. (1957). L'*Hispidae* mineur (*Coelaenomenodera elaeidis* Mlk.), parasite du palmier à huile dans la zone guinéenne, *L'Agronomie Tropicale* 12: 610-632.

Cochereau P. (1965). Notes sur un essai d'acclimatation en Nouvelle–Calédonie de *Tetrastichus brontispae* Ferrière (Hymenopt. Chalc. Eulophidae), *Chrysomelidae Hispinae* sur l'hôte *Brontispa longissima* Gestro var. froggatti Sharp. (Col. Chrysomelidae Hispinae). *Compte Rendu de l' Académie d'Agriculture de France* 19 mai 1965: 661-667.

Cochereau P. (1969). Installation de *Tetrastichus brontispae* Ferr. (Hymenoptera, Eulophidae), parasite de *Brontispa longissima* Gestro, var. *frogatti* Sharp (Coleoptera, Chrysomelidae, Hispinae) dans la presqu'île de Noumea. *Cahier O.R.S.T.O.M., série Biologie,*1969, n° 7: 139-141.

Cohic F. (1961). Polynésie française. Diffusion de deux parasites du cocotier. *Bulletin phytosanitaire FAO* 9 (6): 113.

Corbett G.H. (1923). The two-coloured coconut leaf-beetle. Department of Agriculture. *Straits Settlements and rederated Malay States Bulletin* n° 34.

Cotterell G. S. (1925). The hispid leaf-miner (*Coelaenomenodera elaeidis* Maul.) of oil palms (*Elaeis guineensis* Jacq.) on the Gold Coast. *Bull. Ent. Res.*16: 77-83.

Cox m. L. (1994). The Hymenoptera and Diptera parasitoides of Chrysomelidae. *In:* P.H. Jolivet, M.L. Cox and E. Petitpierre (eds), Novels aspects of the biology of Chrysomelidae. *Dordrecht: Kluwer Academic Publishers*: 419-467.

Dammerman K.W. (1929). The agricultural zoology of the Malay archipelago. Amsterdam 1929.

Dejean A., Djieto-Lordon C., Durand J.L. (1997). Ant mosaic in oil palm plantations of the province of Cameroon: Impact on leaf miner beetle (Coleoptera: Chrysomelidae). *Journal of Economic Entomology* 90 (5): 1092-1096.

Dharmadhikari P.R., Perera P.A.C.R., Hassen T.M.F. (1977). A short account of the biological control of *Promecotheca cumingi* (*Col;: Hispidae*) the coconut leaf-miner, in Sri Lanka. *Entomophaga* 22 (1): 3-18.

Delvare G. (1992). Les Chalcididae d'importance économique dans les palmeraies d'Amérique tropicale (Hymenoptera). *Bulletin de la Société Entomologique de France* 97 (4): 349-372.

Desmier de Chenon R. (1973). Mission entomologique en Colombie et en Equateur. Etude d'*Hispoleptis* et de ses parasites. Rapport de mission non publié, Station de Zoologie INRA Versailles, 66 p.

Ding Siew Ming (1976). Outbreak of *Promecotheca cumingi* baly on coconut palms in Province Wellesley, Malaysia. *Malaysian Agriculture Journal* 1976: 200-220.

Fairmaire L. (1890). Differentes espèces de *Coelaenomenodera*. *Bulletin de la Société Entomologique de France* 26 (11): CCII-CCIII.

Fairmaire L. (1898).– *Coelaenomenodera perrieri*. *Annales de la Société Entomologique Belge* XLII 498.

Fédière G., Monsarrat P., Lecoustre R. (1986). Identification of new viruses of *Coelaenomenodera minuta* (Coleoptera Chrysomelidae) most important issued defoliator oil palm in Ivory Coast. *In* "Extended abstracts of 2nd International Conference on plant protecton in the tropics". Genting Highlands; Malaysia, 17-20 march: 287-288.

Fernando H.E. (1972). Concluding report on control of *Promecotheca cumingi*, the introduced pest of coconut. *Ceylan Coconut Quaterly* 23: 116-122.

Ferreira J. M. S., Morin J.P. (1986). A barata do coqueiro, *Coraliomela brunnea* Thumb. (1981) (Coleoptera: Chrysomelidae). *EMBRAPA. Circular tecnica* n° 1, Aracaju, Brazil, 10 p.

Froggatt J.L. (1936). Coconut pests. *New Guinea Agriculture Gazette* 2:18.

Galego V.C. (1986). Notes on the parasitoid complex of the coconut leafminer (*Promecotheca cumingi* Baly) in some Mindanao Provinces. *Philippine Journal of Coconut Studies* 11 (1): 4-7.

Genty P., Desmier de Chenon R., Morin J.P. (1978). Les ravageurs du palmier à huile en Amérique latine. *Oléagineux* 33 (7): 326-419.

Genty P., Mariau D. (1973). Les ravageurs et maladies du palmier à huile et du cocotier. Le genre *Himatidium*. *Oléagineux* 28 (11): 513-515.

Gestro R. (1908). *Coelaenomenodera*. *Annale delli Museum di Storie Natural Genova* 3 (4): 231-247.

Hargreaves A.R.C.S. (1937). Some insects and their food-plants in Sierra Leone. *Bull. Ent. Res.* 28: 505-520.

Jolivet P. (1997). Biologie des Coléoptères Chrysomelides. Paris, Boubée, 279 p.

Jover H. (1950). Note technique sur la lutte contre *Coelaenomenodera* sp. parasite des palmiers à huile à La Mé. *Oléagineux* 5 (3): 156-160.

Kalshoven L.G.E. (1957). An analysis of ethological, ecological and taxonomic data on oriental hispinae (Coleoptera, Chrysomelidae). *Tijdschrift voor Entomologie Deel* 100, AFL. 1: 5-24.

Kalshoven L.G.E. (1981). The pests of crops in Indonesia PT Ichtiar Baru van Hoeve, Jakarta: 446-458.

Kerrich G.J. (1970). On the taxonomy of some african Eulophidae (Hym. Chalcidoidea) associated with oil palm, coffee and mango. *Bull. Ent. Res.* 60: 327-331.

Kerrich G.J. (1974). Systematic studies on Eulophidae of economic significance (Hymenoptera, Chalcidoidea). *Bull. Ent. Res.* 63: 629-639.

Kowalski J. (1917). Un ennemi du cocotier aux Nouvelles-Hébrides. Le *Promecotheca opacicollis* gestro. *Annales Epiphyties* 4: 286-327.

Laboissière V. (1990). Deux nouveaux *Monolepta* de la collection du musée du Congo. *Revue de Zoologie et de Botanique Africaines* 34 (1): 131-132.

Lecoustre R., Mariau D., Philippe R., Desmier de Chenon R. (1980). Contribution à la mise au point d'une lutte biologique contre *Coelaenomenodera.* II. Introduction en Côte d'Ivoire d'un hyménoptère Eulophidae du genre *Chrysonotomyia* Ashmead, de Madagascar. *Oléagineux* 35 (4) 177-186.

Lepesme P. (1947). Les insectes des palmiers. Paris, Paul Lechevalier: 903 p.

Lever R.J.A.W. (1969). Les ravageurs du cocotier. Etudes agricoles de la FAO n° 77, Rome, 196 p.

Le Verdier A., Genty P. (1988). Dégâts et contrôle de *Hispoleptis subfasciata* Pic. Hispine phyllophage du palmier à huile en Amérique latine. *Oléagineux* 43 (11): 397-407.

Lima M. F. de (1997). Occurrence and damage of *Haemishaerota tristis* (Boheman) (Coleoptera Chrysomelidae), on coconut palms, in the state of Rio de Janeiro, Brazil. *Agrotropica* 9 (3): 141-142.

Liu S.D., Lin S.C. (1989). Microbial control of coconut leaf beetle (*Brontispa longissama)* with green muscaridine fungus, *Metharizium anisopliae* var. *anisopliae. Journal of Invertebrate Pathology* 53 307-314.

Mariau D. (1975). Hispines du genre *Coelaenomenodera* ravageurs du cocotier à Madagascar. *Oléagineux* 30 (7): 303-309.

Mariau D. (1988). The parasitoids of Hispidae in Biology of Chrysomelidae. *In:* P. Jolivet, E. Petitpierre and T.H.H. Siao (eds.) Biology of Chrysomelidae. Dordrecht: Kluwer Academic Publisher: 449-461.

Mariau D. (1999). Les coléoptères *Chrysomelidae* inféodés au palmier à huile et au cocotier et leurs parasitoïdes. *Annales de la Société Entomologique de France* (N.S.) 35 (suppl): 230-237.

Mariau D. (2001). Factors limiting outbreaks of the oil palm pests, *Coelaenomenodera: C. elaeidis* Maulik and *C. lameensis* Berti et Mariau (Coleoptera: Chrysomelidae, Hispinae) in west Africa. Rapport Cirad.

Mariau D., Djob Bikoi J. (1990). *Monolepta apicicornis* Thomson (Coleopt. Chrysomelidae) mineur des racines du palmier à huile en Afrique de l'Ouest. *Oléagineux* 45 (7): 311-317.

Mariau D., Lecoustre R. (2000). Rôle de facteurs éco-climatiques et édaphiques sur la fécondité au champ de *Coelaenomenodera lameensis* mineur des feuilles du palmier à huile en Afrique de l'ouest. *Insect Sci. Applic.* 20 (1): 7-21.

Mariau D., Lecoustre R. (2001).An explanation for outbreaks of *Coelaenomenodera lameensis* Berti and Mariau (Coleoptera: Chrysomelidae: Hispinae), an oil palm (*Elaeis guineensis* Jacq.) leaf miner in west Africa, based on a study of mortality factors. *African Entomology* (sous presse).

Mariau D., Morin J.P. (1971). La biologie de *Coelaenomenodera elaeidis* Mlk. II Description et biologie des principaux parasites. *Oléagineux* 26 (2): 83-88.

Mariau D., Morin J.P. (1972). La biologie de *Coelaenomenodera elaeidis*. IV. La dynamique des populations du ravageur et de ses parasites. *Oléagineux* 27 (10): 469-474.

Mariau D., Morin J.P. (1974). La biologie de *Coelaenomenodera elaeidis* Mlk. VI La mortalité naturelle des larves. *Oléagineux* 29 (12): 549-555.

Mariau D., Philippe R. (1983). Avantages et inconvénients des méthodes de lutte chimique contre *Coelaenomenodera minuta* (Coleoptera, Chrysomelidae), Hispine mineur du palmier à huile; *Oléagineux* 38 (6): 365-370.

Mariau D. Philippe R., Lecoustre R., (1978). Les parasites larvaires de *Coelaenomenodera elaeidis* Mlk, hispine du palmier à huile en Afrique occidentale. *Oléagineux* 33 (4): 153-160.

Maulik S. (1920). A new hispid beetle injurious to the oil palm in the Gold Coast. *Bull. Ent. Res.* 10 (2): 171-174.

Maulik S. (1937). Distributional correlation between Hispin beetles and their host-plants. *Proceeding Revue Zoology Society, London*: 669.

Maulik S. (1940). The food plant of *Platyauchenia latreillei. Annals and Magazine of Natural History* 5 (2): 256.

Morin J.P., Mariau D. (1970). Etude sur la biologie de *Coelaenomenodera elaeidis.* I Morphologie et étude du développement. *Oléagineux* 25 (1): 11-16.

Morin J.P., Mariau D. (1971). La biologie de *Coelaenomenodera elaeidis* Mlk. III. La reproduction. *Oléagineux* 26 (6): 373-378.

Morin J.P., Mariau D. (1974). La biologie de *Coelaenomenodera elaeidis* Mlk. V. La mortalité au niveau des oeufs. *Oléagineux* 29 (5): 233-238.

Moura J.I.L. (1985). Incidencia de *Calyptocephala brevicornis* (Boheman 1850) (Coleoptera: Chrysomelidae Cassidinae) en cultura de dende (*Elaeis guineensis*) no municipio de Porto Platon Territorio federal do Amapa. *Anais da sociedade entomologica do Brasil*, 14 n° 1: 37-43.

Muniappan R., Duenas J.G., Blas T.(1980). Biological control of the Palau coconut beetle, *Brontispa palauensis* (Esaki and Chupo), on Guam. *Journal of the University of Guam* 16 (2): 359-360.

Nair M.R.G.K., Oommen C.N. (1965). *Callispa* sp. (Hispinae: Chrysomelidae) as a minor pest of coconut in Kerala. *Indian J. Ent.* 27 (2): 230-232.

Perera P.A.C.R. (1982). Coconut pests in Sri Lanka – the coconut leaf-miner. *Planter* 58: 157-162.

Philippe R. (1977). Etude du développemen de *Coelaenomenodera elaeidis* Mlk. *(Coleoptera, Hispidae)* sur hybride *E. guineensis* × *E. melanococca. Oléagineux* 32 (1): 1-4.

Philippe R. (1990a). Modalités pratiques de lutte chimique contre *Coelaenomenodera minuta* Uhmann (Cleoptera-Chrysomelidae-Hispinae). *Oléagineux* 45 (4): 165-173.

Philippe R. (1990b). Etude de l'action de l'Evisect S sur *Coelaenomenodera minuta* (*Coleoptera-Chrysomelidae-Hispinae). Oléagineux* 45 (4): 143-163.

Philippe R., Desmier de Chenon R., Lecoustre R., Mariau D. (1979). Contribution à la mise au point d'une lutte biologique contre *Coelaenomenodera:* Introduction en Côte d'Ivoire de parasites larvaires d'hispines. *Oléagineux* 34 (6): 271-279.

Philippe R., Mariau D. (1996). Organismes entomopathogènes. *In:* D. Mariau Lutte intégrée contre les ravageurs des cultures perennes tropicales. Cirad: 113-132.

Quicke D.L.J., Polaszek A. (2000). A new genus, and first host records, for the Adeshini: parasitoids of hispine beetles (Braconidae: Braconinae; Coleoptera: Chrysomelidae). *Journal of Hymenoptera Research* 9 (1): 104-107.

Risbec J. (1934). Le cocotier en Nouvelle Calédonie. *Agronomie Coloniale* n° 199-200.

Sandino E.V. (1972).– *Cephaloleia* sp. *cerca vagelineata* Pic una plaga de la palma africana. *Facultad National Agronomia Antioquia* 27 (2): 75-77.

Santos J.V. (1968). Algunas caracteristicas biologicas y etiologicas del *Alurnus humeralis* Rosenberg "gusano chato o cogollero" de la palma africana.*Oléagineux* 23 (3): 159-164.

Stapley J.H. (1980). Coconut leaf beetle (Brontipa) in the Solomons. *Alafua Agricultural Bulletin* 5 (4): 17-22.

Taylor T.H.C. (1937). The biological control of an insect in Fiji. *London, Imperial Institute of Entomology.*

Uhmann E. (1957). Coleopterum catalogus. Chrysomelidae Hispinae, Dr.W. Junk ed. Netherlands, 398 p.

Uhmann E. (1963). Hispinae (Cryptonychini) von den Mascarenen. *The Mauritius Institute Bull* 5 (7): 287-298.

Viggiani G. (1980). Two new species of *Oligosita* Walk. (*Trichogrammatidae*), egg-parasites of Hispids (Coleoptera) in Madagascar. *Bolletino del laboratorio di entomologia agraria "Filippo Silvestri" di Portici* 37: 3-7.

Voegele J. M. (1989). Biological control of *Bronstipa longissima* in Western Samoa: an ecological and economic evaluation. *Agriculture, ecosystems and environment* 27: 315-329.

Waterston J. (1925). On some Eulophid parasites (Hym., Chalcidoidae) of the oil palm hispid beetle. *Bull. Ent. Res.* 15 (4): 313-400.

Zenner J.I. (1968). *Delocrania cossyphoides* Guer., plaga potential del cocotero en Colombia. *Agric. Trop., Bogota* 24: 202-203.

CURCULIONIDAE

The family of the *Curculionidae* is extremely abundant in species, and a very large number of them is found on oil palm and coconut. However, few of the species are major pests and some of them are even very useful in *Elaeis* reproduction.
Lepesme (1947) inventoried 12 subfamilies, though some of them are not found on oil palm and coconut. The family was revised by Thompson (1992).

BARIDINAE

The species in this subfamily usually live off flowers. Such is the case with ***Microstrates ypsilon*** Marshall and ***M. bondari*** Hustache, ***Pseudocentrinus lucidulus*** Hustache and ***Hustachea (Tonesia) bondari*** (Hustache), on male coconut flowers in Brazil (Bondar, 1940). Many other species belonging to these genera can be found on the male flowers of other palms. We would also mention ***Tonesia melas*** Boheman, which is more frequent on rachis wounds on *Cocos* in Brazil (Bondar 1940), though we found it under the same conditions on *Elaeis* in Colombia, ***Catollerus strangularis*** Hustache (Hustache *in* Lepesme 1947) and ***Dialomia polyphaga*** Bondar (***D. discreta*** Casey) (Bondar 1940), both on *Cocos* spathes in Brazil.
The species of the genus ***Parisoschoenus*** attack fruits, such as ***P. obesulus*** Casey which would seem to be responsible for the fall of numerous young fruits (Bondar 1940) and ***P. suturalis*** Casey, which is also found on *Elaeis*.

ERIRHININAE

Two of the 60 species reported on the flowers of various palms by Lepesme in Brazil have been observed on *Elaeis* or *Cocos*. We detected ***Celestes*** on *E. oleifera* flowers in Brazil, whereas ***Ancylorrhynchus trapezicollis*** Hustache and ***Balanephagus attaleae*** Hustache are found on *Cocos* (Bondar 1940). ***Nodocnemus uniformis*** Marshall develops on male *Cocos* flowers in the Pacific region (Risbec 1937), ***Amorphoidea*** sp. in Malaysia (Miller *in* Lepesme 1947) and ***A. coimbatorensis*** Subramaniam in India (Subramaniam *et al.* 1975). A species of the same genus attacks the base of spikelets in Sri Lanka (unpublished).

CRYPTORRHYNCHINAE

A single species, ***Palaecorynus cerinipes*** Hustache, is very common in the Congo on *Elaeis* fronds (Ghesquière *in* Lepesme 1947).

DERELOMINAE

This subfamily contains a large number of species, all of which live off *Cocos* or *Elaeis* flowers. ***Meredolus cocois*** Marshall in the Solomon Islands (Lever 1937) and ***Derelomorphus eburneus*** Marshall in Malaysia (Corbett 1932) have been found on male *Cocos* flowers. A species of ***Meredolus*** has been identified in Sri Lanka, where it causes desiccation of leaves and inflorescences (Mahindapala and Subasinghe 1979).
Notolomus, N. elaeisae Hustache and ***N. mariae-helenae*** Bondar have been observed in Brazil on male *Elaeis* and *Cocos* flowers (Bondar 1941).

Prosoestus minor Marshall and ***P. sculptilis*** Faust attack *Elaeis* flowers in West Africa (Philippe 1993) in which the female lays its eggs. The damage caused by ***P. minor*** is only slight, since the larva gallery stops at the base of the style. With ***P. sculptilis*** the larva gallery reaches the gynaecium and the flower is destroyed.

The species of the genus ***Elaeidobius*** were once included under the previous genus. All the species belonging to this genus lay their eggs in male *Elaeis* flowers in Africa. The male flower is destroyed but pollen emission is not disrupted. Five species have been described: ***E. kamerunicus*** (Faust), ***E. subvittatus*** (Faust), ***E. plagiatus*** (Fabricius), ***E. singularis*** (Faust) and ***E. bilineatus*** (Hustache), the last two are much less numerous than the first two (Desmier de Chenon 1981, Mariau *et al.* 1991).
The role played by these insects in pollination was suspected as early as 1910 by Chevalier, an idea countered by Gater, and taken up again by Henry (1948), then once again abandoned. It was only at the end of the 70s that the idea was taken up again due to the poor pollination of oil palm in Malaysia (Gray 1966, Hardon 1973), which necessitated assisted pollination. As ***E. kamerunicus*** was a better pollinator than the other species (Syed 1979), it was introduced into Malaysia (Syed 1982) after in-depth quarantine studies (Kang and Abdul Karim 1982), then in Indonesia. It was then introduced a little later in tropical American countries (Mariau and Genty 1988) where pollination was ensured in many places by ***E. subvittatus***, whose larvae can be attacked by *Stodiplosis gagnei* Baylac (1987) (Dipt. *Cecidomyiidae*), which had previously been introduced at an unknown date, as well as by a species in the family of the *Nitidulidae* and of the genus ***Mystrops*** (see this family) (Genty *et al.* 1986). Lastly, ***E. kamerunicus*** was released in Madagascar in 1993 (at the Antalaha estate, Mariau unpublished) where another *Nitidulidae,* but a very poor pollinator, already existed. All these introductions were considerably successful everywhere, and have been reported in numerous publications. In 1983, i.e. about a year after introduction,

there was a substantial increase in yields in Malaysia linked to an increase in average bunch weight.

MOLYTINAE

Homalinotus coriaceus Gyllenhall (Bondar 1940, Fonseca 1962, Vaurie 1973)

The adult is black and measures up to 25 mm in length. The female lays 1 or 2 eggs at the base of a flower stalk on coconut palms in Brazil, in which the larvae mine a gallery, with a small surface excavation in the stem. On young inflorescences, such attacks can lead to rachis rot which, on older inflorescences, can cause the rachis to snap if the bunch contains many nuts (Mariau 1985). The development cycle is estimated at 6-8 months (Bondar 1940). The morphology and physiology of the reproductive system in this species have been studied in depth (Nascimento and Carvalho 1970). Studies have also been made of variations in oxygen consumption (Nascimento 1973).
Several other species are mentioned, ***H. porosus*** Gyllenhall and ***H. deplanatus*** Shalb. (Bondar 1940), but also ***H. calvescens*** Dohrn (Moreira *in* Lepesme 1947).

Amerris (***Amerrhinus***) ***ynca*** (Sahlberg) (Bondar 1940, Moura *et al.* 1994)

The adult is pale yellow and measures 26 mm in length; the body is covered in black protuberances. The cycle lasts around 8 months (Moura *et al.* 1994). The adult lays its eggs at the base of the rachis of young fronds. The larva gallery mined into the rachis causes the frond to dry out and snap.
A. parcus Germ. (***Cholus parcus***) has also been observed on coconut in Brazil.
C. zonatus Swed shows the same behaviour as ***A. ynca*** in Saint Lucia (Parasram and Mederick 1971).

RHYNCHOPHORINAE

Rhynchophorus

In the revision of this genus, Wattanapongsiri (1966) mentions 9 species on *Elaeis* and *Cocos:* in Latin America and the Caribbean, ***R. palmarum*** (L.) and ***R. ritcheri*** Watta; in the Asia-Pacific zone ***R. ferrugineus*** (Oliv.), ***R. bilineatus*** (Montr.), ***R. distinctus*** Watta., ***R. lobatus*** Rits, and ***R. vulneratus*** (Panzer); ***R. ferrugineus*** was reported in Egypt in 1992 on date palm by El Garhy (1996), then in the United Arab Emirates and southern Spain on ornamental palms

(Barranco *et al.* 1996). Lastly, ***R. phoenicis*** (Fabr.) and ***R. quadrangalus*** Quedenfeldt in Africa.

Description

This concerns only the 5 main species. The adults are weevils measuring 25 to 50 mm in length, depending on the species and the individuals. The head is extended by a long rostrum which, in males, has bristles on the upper distal section. ***R. palmarum*** is uniformly black, ***R. phoenicis*** is blackish-brown with two dark red stripes on the thorax and elytra, ***R. ferrugineus*** is orangey-red, with black stripes on the elytra and black patches on the thorax, ***R. vulneratus*** is black with often faint orangey longitudinal stripes on the thorax.
The larvae of all these species resemble each other. They range from 40 to 60 mm in the final instar and are whitish, apodal and swollen in the median section. The brown head has powerful mandibles.
The pupa is enclosed in a fibrous cocoon.

Biology (Wilson 1962, 1963, Sivapragasam *et al.* 1990, Nirula 1956, Mexzon *et al.* 1994, Hagley 1964, Goonewardena et Velu 1958, Mariau *et al.* 1981, 1991, Genty *et al.* 1978, Pillai 1984, Thajudin 2000, Jaya *et al.* 2000).

The larval cycle (7-8 instars) lasts 2 to 3 months, under rearing conditions, with variations depending on the species and the diet. Fecundity exceeds 200 eggs per female, with an oviposition period of more than 1 month. For research purposes, larvae can usually be reared in the laboratory on sugar cane (Rahalkar *et al.* 1972, Ranavare *et al.* 1975), and the development cycle takes longer on that substrate than on oil palm (Zagatti *et al.*1993). Rearing can also be carried out on an artificial medium (Rahalkar *et al.* 1978).

Rhynchophorus populations are attacked by very few parasitoids or predators. *Scolia erratica* Smith (Hym. *Scoliidae*) can attack ***R. ferrugineus*** (Burkill 1917). The pupa of that species can be attacked and killed by mites (Nirula 1956). Over 50% of ***R. palmarum*** larvae can be parasitized by the tachinid *Paratheresia menezesi* Townsend in Brazil (Moura *et al.*, 1993). Another tachinid, *Parabillea rhynchophorae* Blanch has been reported in Bolivia (Candia and Simmonds 1965). The same *Rhynchophorus* is attacked by *Tetrapolipus rhynchophori* Ewing (*Acarina Pyemotidae*) (Nirula 1956). The nematode *Rhabditidae Terathorabditis palmarum* Gerber and Giblin-Davis (1990a) has been observed in a ***R. palmarum*** cocoon, but the effect on insect development is unknown. Lastly, a cytoplasmic polyhedrosis virus has been isolated from the larvae of ***R. ferrugineus*** but the incidence of this virus on populations is also unknown (Gopinadhan *et al.* 1990).

The larvae develop in the living tissue of palms to which they gain access via a wound. Maharaj (1962) believed that the insect could penetrate healthy palms, but that has never been reported by any other authors. If the larvae bore into the tissues in the upper crown, the palm often dies after the attack. ***R. palmarum*** is considered to be the main vector of red ring disease, the biological agent being the nematode *Rhadinaphelenchus cocophilus* (Cobb) (Goodey 1960, Dean and Velis 1976, Gerber and Giblin-Davis 1990b, Gerber *et al.* 1991, Toquica 1993, Hagley 1963, Griffith, 1968); small adults are more contaminated than the others (Griffith 1974). Some ***R. palmarum*** adults use an enzyme to eliminate injected nematodes (Griffith 1976). It has been possible for nematodes to penetrate the plant through healthy tissues when contaminated adults were placed in leaf axils (Morin, personal communication). Fifteen types of bacteria have been isolated from ***R. ferrugineus*** larvae, 5 of which are potentially pathogenic (ICAR 1992).

Chemical ecology

As adults are attracted by the smell of fresh oil palm tissues, the attraction has been used to trap adults (Maharaj 1973, Morin *et al.* 1986, Rochat 1990). Studies were carried out to study which molecules were attractive (Hagley 1965, Fandialan and Fernandez 1983, Gunatilake and Gunawardena 1986), Nagnan *et al.* 1992, Gouraud 1994, Gries *et al.* 1994, Gunawardena 1994a,b, Gunawardena *et al.* 1998) to obtain a composition that was just as attractive, when combined with the pheromone (see below), as oil palm tissues themselves (Rochat *et al.* 2000).

Meksongsee and Sakulpanich (1975) were first to put forward the hypothesis that ***R. vulneratus*** adults might emit a pheromone. Research was only carried out later and the first aggregation pheromone was discovered in ***R. palmarum*** (Rochat 1991, Rochat *et al.* 1991a,b, 1993a, Chinchilla *et al.* 1993, Jaffé *et al.* 1993, Oehlschlager *et al.* 1992, 1993a,b,c, 1995a), a species in which a cuticular pheromone, whose role was unknown, was also detected (Rochat 1991). That discovery led to numerous studies on the other species of ***Rhynchophorus*** in both the laboratory and the field, leading to the detection of an aggregation pheromone in the main species: ***R. phoenicis*** (Gries *et al.* 1993, Mori *et al.* 1993a, Rochat *et al.* 1993b, 1995, Perez *et al.* 1994), ***R. ferrugineus*** and ***R. vulneratus*** (Hallett *et al.* 1993a, 1993b, Mori *et al.* 1993b, El Garhy 1996, Perez *et al.* 1996), and ***R. bilineatus*** (Oehlschlager *et al.* 1995b). These pheromone secretions usually contain several compounds, one of which is a majority compound. The three Asian species studied have the same majority compound, (4*S*,5*S*)-4 methyl-5-nonanol, whereas that of the African and American species is (3*S*,4*S*)-3-methyl-4-octanol for ***R. phoenicis*** and (4*S*,2*E*)-2-methyl-5-

hepten-4-ol for ***R. palmarum*** (Rochat *et al.* 1991b, Oehlschlager *et al.* 1992).

Control methods

Control of the different ***Rhynchophorus*** species is primarily preventive, by halting primary attacks linked to other insects, rodents, etc (Mariau 1968). For curative action, insecticide applications can be tried (Abraham *et al.* 1975, Ganeswara Rao *et al.* 1989, Sadakathulla 1991). Attempts have been made to apply the male sterilization method to ***R. ferrugineus*** (Kloft *et al.* 1986), apparently without success.

As shown by earlier studies, the most effective method, apart from preventing wounds, is trapping with traps containing the pheromone and a piece of plant matter (various palms, sugar cane) or a combination of allelochemical substances (Oehlschlager *et al.* 1993a, 1995a, Hallet *et al.* 1993b, 1999) apparently without success.

In the case of Red Ring disease, which is primarily transmitted by *R. palmarum* in Latin America, trapping with pheromone has been found to play a part in increasing insect movements over long distances, increasing the possibilities of nematode transmission to healthy palms (Genty personal communication).

Dynamis borassi Fabricius

This insect, which is found on coconut inflorescences in tropical America (Giblin-Davis *et al.* 1995c), is morphologically very similar to ***R. palmarum***. However, damage has never been reported on palms. It emits a pheromone, whose majority compound is the same as that of the Asian species, hence different from that of ***R. palmarum*** (Giblin-Davis *et al.* 1997).

Metamasius

This genus comprises numerous species (Vaurie 1966). Only the species ***M. hemipterus*** L. has been reported on *Elaeis* and *Cocos,* but it mostly develops on sugar cane (Restrepo *et al.* 1982). The development cycle for the 7 to 9 larval instars lasts 2 months. This species is mainly found on the cross-sections of fronds shortly after they are cut, the larvae developing in the part of the rachis remaining on the palm, but they remain on the surface and do not therefore damage the plant. ***M. hemipterus*** is sometimes mentioned as being a vector of red ring disease. This species emits a pheromone whose majority compound is the same as that of ***Dynamis*** and of the Asian species of ***Rhynchophorus*** (Ramirez-Lucas *et al.* 1996a,b, Perez *et al.* 1997). Trapping techniques have also been developed for ***M. hemipterus*** (Giblin-Davis *et al.* 1995a), and the use of chemical pesticides and an entomopathogenic nematode has been studied

(Giblin-Davis *et al.* 1995b). Lepesme (1947) mentions other species on *Cocos*.

Temnoschoita quadripustulata Gyllenhall

The adult of this African species does not exceed 12 mm in length and is black. The deeply grooved elytra have rust-coloured patches.
The development cycle lasts around 2 months and it has the same behaviour as ***R. phoenicis*** but this species only attacks nursery plants in the first two years (Mariau 1974, Mariau *et al.* 1981). Attacks usually take place in poorly aerated plantings (shading in the nursery, intercropped planting under old palms). In adult plantings, the insect can develop on the cross-sections of fronds like ***Metamasius*** in America. Five other species have been reported mostly on *Elaeis* and sometimes on *Cocos*: ***T.maynei*** Hustache, ***T. subulirostris*** Kolbe, ***T. erudita*** Duvivier, ***T. basipennis*** Duvivier and ***T. delumbata*** Boheman. These species show the same behaviour as ***T. quadripustulata*** (Conzotte and Ghesquière *in* Lepesme 1947, Jerath 1965) but no serious attacks have been reported for these species.

Rhinostomus barbirostris Fabricius (Bondar 1940, Vaurie 1970, Ferreira and Morin 1984)

The adult is blackish-brown and varies considerably in size (15 to 45 mm). The rostrum of the male bears long dense hairs.
The female lays its eggs in coconut stems, and the larvae bore into the stem, which can be almost completely destroyed. The complete development cycle lasts around 6 months. Serious attacks have only been reported in Brazil. ***R. scrutator*** Ol. has been reported in Cuba (Wolcott 1933). The tachinid *Paratheresia brasilensis* Tns. has been isolated from ***R. barbirostris*** in Trinidad (Candia and Simmonds 1965). Research on its sexual behaviour has been carried out by Eberhard (1983). The adult is a vector or red ring disease (Dean and Velis 1976).

R. afzelii Fåhraeus exists in Africa. In West Africa, we have only seen it on dead coconut palms. ***R. niger*** Druzy must behave in the same way in Madagascar and on Ngazidja in the Comoro Island (Lepesme 1947).

Rhabdoscelus obscurus Boisduval (Bianchi et Owen 1965, Dharmaraju *et al.* 1980)

It mostly attacks sugar cane, and other plants secondarily, including coconut. It is widespread from the Moluccas to the islands of the Pacific and in Australia (Queensland). In Vanuatu, it has been seen mining galleries in the petiole bases and stem.

In Papua New Guinea, the insect is parasitized by *Ceromasia sphenophori* Villeneuve (*Diptera Tachinidae*) and a *Braconidae*.
We would also mention two other species: ***R. asperipennis*** (Fairmaire), which may be mixed with the former (Bianchi and Owen 1965) and ***R. lineatocollis*** Hell, which is reported in the Philippines (Herms 1926, Gabriel 1976).

Eugnoristus monachus Ol. var. **alluaudi** Hustache (Frappa 1933). The larvae of this species bore into frond petioles in Madagascar.

Sparganobasis subcruciatus Marshall attacks coconut stems in Papua New Guinea, which can subsequently snap (Froggatt 1936, Simmonds 1938).

Myllocerus curvicornis (F.) (Kurian *et al.* 1978). The larvae attack coconut fronds in India. Attacks on coconut palms vary depending on the mineral nutrition of the palms (Ponnamma *et al.* 1982).

Diocalandra taiitensis Guerin
The adult is black with dark red patches on the thorax and elytra. It measures 6 to 7 mm in length. The development cycle lasts 10 to 12 weeks (Herms 1926). The larvae can bore into all the organs of the plant: stem, fronds, fruits, roots (Lever 1941). In Madagascar, the larvae only attack the stem (Frappa 1947). It causes serious damage in New Caledonia (Risbec 1936).

Diocalandra frumenti Fabricius
This species is widespread in Asia (from India to the Pacific) (Corbett 1932, Purohit *et al.* 1988, Dharmaraju 1983). Damage is similar to that caused by ***D. taiitensis***. This species was previously called ***D. stigmaticollis*** (Louis and Chandarasekaran 1980). Attacks lead to a drop in yields and can kill coconut palms. 40% of larvae can be attacked by the Hymenoptera *Braconidae Spathius apicalis* Westwood and the predators *Plaesius javanus* Erichson (Col. *Histeridae*) and *Chrysophilus ferruginosus* Wiedemann (Dipt. *Rhagionidae*) (Lever 1969).
D. impressicolis Quedenfeldt is reported on *Elaeis* bunches in central Africa (Ghesquière *in* Lepesme 1947) and ***D. stigmaticollis*** on coconut in Tamul Nadu (India) (Louis and Chandrasekaran 1980).

Cydostethus sp. larvae have been observed in coconut stems in Vanuatu (Kuschel, personal communication).
Lastly, a few species live in dead *Cocos* wood: ***Protocerius colossus*** Ol., ***Omotemnus miniatocrinitus*** Chav. and ***Cyrtotrachelus*** sp. in the Sunda islands (Preuss *in* Lepesme 1947).

COSSONINAE

The species of this subfamily are polyphagous and attack rotten wood. Such is the case with ***Stenotrupis filum*** Fremery on coconut in Samoa (Swezey *in* Lepesme 1947), ***Leurostenus elaeidis*** Marshall on *Elaeis* in the Congo (Ghesquière *in* Lepesme 1947), ***Oxydema fusiforme*** Woll. on *Cocos* in Indonesia (Leefmam *in* Lepesme 1947) and ***Aphanocorynes humeralis*** Marshall on *Cocos* in the Pacific (Lepesme 1947).

CURCULIONIDAE REFERENCES

Abraham V.A., Abdulla Koya K.M., Kurian C. (1975). Evaluation of seven insecticides for Control of red palm weevil *Rhynchophorus ferrugineus* Fabr. *Journal of Plantation Crops* 3 (2): 71-72.

Barranco P., Cabello T. (1996). El picudo rojo de las palmeras *Rhynchophorus ferrugineus* (Olivier), nueva plaga en Europa. (Coleoptera, Curculionidae). *Phytoma* n° 76: 36-40.

Baylac. M. (1987). Description d'un *Lestodiplosis* nouveau, prédateur *d'Elaeidobius subvittatus* (Col. Curculionidae) pollinisateur du palmier à huile *Elaeis guineensis* en Colombie (Dipt. Cecidomyiidae). *Bull. Soc. Ent. Fr.* 92 (3-4): 125-132.

Bianchi F.A., Owen R.P. (1965). Observations on *Rhabdoscelus obscurus* (Boisduval) and *Rhabdoscelus asperipennis* (Fairmaire) in Micronesia (Coleoptera: Curculionidae). *Proceedings, Hawaïan Entomological Society* 19 (1): 77-81.

Bondar G. (1940) Insectos nocivos et molestias do coqueiro. *Bol. Inst. Centr. Fom. Econ. Bahia* n° 8.

Bondar G. (1941). Palmeiras do genero Cocos e descriçào de duas especies novas. *Bol. Inst. Centr. Fom. Econ. Bahia n° 9.*

Burkill J.H. (1917). *Scolia erratica* Smith, a parasite of the red coconut weevil, *Rhynchophorus ferrugineus. Gardens Bull. Straits Settlements* 1 (11-12): 399.

Candia J.D. Simmonds F.J. (1965). A tachinid parasite of the palm weevil, *Rhynchophorus palmarum* L. *in:* Bolivia. *Techn. Bull. Commonw. Inst. Biol. Control* 5: 127-128.

Chevalier A. (1910). Documents sur le palmier à huile: 71-72 *in:* Les végétaux utiles de l'Afrique tropicale française Vol. 7. Paris, 127 p.

Chinchilla C.M., Oehlschlager A.C., Gonzalez L.M. (1993). Use of pheromone-based traps for the management of the American palm weevil, *Rhynchophorus palmarum* in a commercial oil palm plantation, in Proceedings, International Oil Palm Congress, 20-25 September 1993. Palm Oil Research Institute of Malaysia, Kuala Lumpur, Malaysia.

Corbett G.H. (1932). Insects of coconut in Malaya. *Dept. Agri. SSFMS Gen. Ser.* n° 10.

Dean C.G., Velis M. (1976). Differences in the effects of Red Ring Disease on coconut palms in Central America and Caribbean and its control. *Oléagineux* 31 (7): 321-326.

Desmier de Chenon R. (1981). New data on the entomophil pollination of oil palm in West Africa. *Conf. Tech. Oil Palm* AP 11 Kuala Lumpur, 45 p.

Dharmaraju E. *et al.* (1980). The sugar cane weevil on coconut in western Samoa. *Alafua agric. Bull. West Samoa* 4 (2): 8-9.

Dharmaraju E. (1983). First record of the red brown weevil, Diocalandra taïtensis Guerin-Meneville, as a pest of the coconut palm in Tokelau. *Alafua Agric. Bull. West Samoa* 8 (1): 52.

Eberhard W. G.(1983). Behavior of adult bottle brush weevils (Rhinostomus barbirostris) (Coleoptera: Curculionidae) *Rev. Biol. Trop.* 31 (2): 233-244.

El Garhy M.E. (1996). Field evaluation of the aggregation pheromone of the red palm weevil, *Rhynchophorus ferrugineus*, in Egypt. Brighton Crop Protection Conf. Pests and Diseases. 1996.

Fandialan M.M., Fernandez W.L. (1983). Chemical changes in the Lambanog toddy. *Philippines J. Coconut Studies* 8 (1-2): 13-17.

Ferreira J.M.S, Morin J.P. (1984). Estudo sobre o desenvolvimento da broca do tranco do coquiero *Rhinostomus barbirostris* Fab. (Coleoptera: Curculionidae). *Pesquisa em Andamento (EMBRAPA, UEAPE/Aracaju)* n 26, 4 p.

Fonseca J.P. (1962). Broca dos pedunculos dos coqueiros *(Homalinotus coriaceus* Gyllenhal 1836). *O Biológico* 20: 20-25.

Frappa C; (1933). Un nouveau parasite du cocotier à Madagascar. *Eugnoristus monachus* Ol. var. Alluandi Hustache. *Agronomie Coloniale* n° 190.

Frappa G. (1947). Un nouveau parasite du cocotier à Madagascar. *Agro. Trop. 2 (5-6): 299-302.*

Froggatt J.L. (1936) Coconut pest. new Guinea. *Agr.Gaz.* II: 18.

Gabriel B.P. (1976). Pests of coconut in the Philippines. *Philippine J. Coconut studies* 1 (1): 15-25.

Ganneswara Rao A., Ramamohana Rao P., Ramamohana Rao T., Lakshminarayana K. (1989). Studies of the effect of root feeding of systemic insecticides in the control of red palm weevil, *Rhynchophorus ferrugineus* Fab. in coconut. *Indian Coconut J.* 21 (1): 12-16.

Genty P., Desmier de Chenon R., Morin J.P. (1978). Les ravageurs du palmier à huile en Amérique latine. *Oléagineux* 33 (7): 326-419.

Genty P., Garzon A., Lucchini F., Delvare G. (1986). Polinización entomofila de la palma africana en American tropica. *Oléagineux* 41 (3): 99-112.

Gerber K., Giblin-Davis R.M. (1990a) *Teratorhabiditis palmarum* n.sp. (Nemata: Rhabditidae):An Associate of *Rhynchophorus palmarum* and *R. cruentatus. Journal of Nematology* 22 (3): 337-347.

Gerber K., Giblin-Davis R.M. (1990b). Association of the red ring nematode, *Rhadinaphelenchus cocophilus*, and other nematode species with *Rhynchophorus palmarum* (Coleoptera: Curculionidae). *J. Nematol.* 22: 143-149.

Gerber K., Giblin-Davis R.M., Escorbar-Goyes J. (1991). Association of the red ring nematode *Rhadinaphelenchus ocophilus*, with weevils from Ecuador and Trinidad. *Nematropica* 20: 39-49.

Giblin-Davis R.M., Gries R., Gries G., Peña-Rojas, E. Pinzón I., Peña J.E., Perez A.L., Pierce H.D.Jr., Oehlschlager A.C. (1997). Aggregation pheromone of palm weevil *Dynamis borassi. J. Chem. Ecol.* 23: 2287-2297.

Giblin-Davis R.M., Oehlschalger A.C., Perez A.L., Gries G., Gries R., Weissling T.J., Chinchilla C.M., Peña J.E., Hallett R.H., Pierce Jr. H.D., Gonzalez L.M. (1995c). Chemical and behavioral ecology of palms weevils (Curculionidae: Rhyncophorinae) Behavioral Ecology Symposium. *Florida Entomologist* 79 (2): 153-167.

Giblin-Davis R.M., Peña J.E., Duncan R.E. (1995b). Evaluation of an entomopathogenic nematode and chemical insecticides for control of *Metamasius hemipterus sericeus* (Coleoptera: Curculionidae). *J. Entomol. Sci.* 31.

Giblin-Davis R.M., Peña J.E., Oehlschalger A.C., Perez A.L. (1995a). Optimization of semiochemical-based trapping of *Metamasius hemipterus sericeus* (Olivier) (Coleoptera: Curculionidae). *J. Chem. Ecol.* 22: 1389-1410.

Goodey J.B. (1960). *Rhadinaphelencus cocophilus (*Cobbe 1919) n. comb., the nematode associated with “ red-ring ” disease of coconut. *Nematologica* 5 (2): 98-102.

Goonewardena H.F., Velu M.S. (1958). The red palm weevil *(Rhynchophorus ferrugineus* Ol.) in Ceylon Part I Introduction, distribution and the life history. *Ceylon Coconut Quarterly* 9 (1/2): 20-40.

Gopinadhan P.B., Mohandas N., Nair K.P.V. (1990). Cytoplamic polyhedrosis virus infecting red plam weevil of coconut. *Current Science* 59 (11): 577-580.

Gouraud P. (1994). Contribution à l’identification de synergistes végétaux de la phéromone du charançon des palmiers (Coleoptera: Curculionidae). Mémoire présenté pour l’obtention d’un D.E.A. Université Paris VI. INRA Versailles.

Gray B.S. (1966) The necessity for assisted pollination in areas of low male inflorescence production ant its effect on the component of yield of the oil palm *(Elaeis guineensis*). *Planter* 42: 16-22.

Gries G., Gries R., Perez A.L., Gonzales L.M., Pierce H.D.Jr., Oehlschlager A.C.,Rhainds M., Zebeyou M., Kouame B. (1994). Ethyl propionate: synergistic kairomone for African palm weevil, *Rhynchophorus phoenicis* L. (Coleoptera: Curculionidae). *J. Chem. Ecol.* 20: 889-897.

Gries G., Gries R., Perez A.L., Oehlschlager A.C., Gonzales L.M., Pierce H.D.Jr., Kouda M., Zebeyou M., Nanou N. (1993). Aggregation pheromone of the African palm weevil, *Rhynchophorus phoenicis* F. *Naturwissenschaften* 80: 90-91.

Griffith R. (1968). The mechanism of transmission of the red ring nematode. *Jour. Agric. Soc. Trinidad and Tobago* 4. 1968.

Griffith R. (1974) The use of the smaller palm weevil, *Rhynchophorus palmarum* L., in the forcasting of red ring disease outbreaks. *J. Agric. Soc. Trinidad and Tobago* 74 (2): 149-158.

Griffith R. (1976). Some characteristics of a defense mechanism in the palm weevil, *Rhynchoporus palmarum* L., against the red ring nematode *Rhadinaphelencus cocophilus* (Cobb) Goodey 1960. *J. Agric. Soc. Trinidad Tobago* 76 (3): 246-258.

Gunatilake R., Gunawardena N.E. (1986). Ethyl alcohol: a major attractant of red weevil *Rhynchophorus ferrugineus* F. *Proc. Sri Lanka Assoc. Adv. Sci. 42nd Ann. Sess.*: 70.

Gunawardena N.E. (1994a). Terpenes as potentiel semiochemicals for the coconut pest, *Rhynchophorus ferrugineus* (Coleoptera: Curculionidae): an electroantennogram assay. *J. Nat. Sci. Coun. Sri Lanka* 22 (1): 35-42.

Gunawardena N.E. (1994b). Steam volatiles of coconut bark: Chemical investigations and electroantennogram responses of the coconut pest, *Rhynchophorus ferrugineus* (Coleoptera: Curculionidae). *J. Nat. Sci. Coun. Sri Lanka* 22 (3): 231-238.

Gunawardena N.E., Kern F., Janssen E., Meegoda C., Schäfer D., Vostrowsky O., Bestmann H.J. (1998). Host attractants for red weevil, *Rhynchophorus ferrugineus*: identification, electro physiological activity, and laboratory bioassay. *Journal of Chemical Ecology* 24 (3): 425-437.

Hagley E.C. (1963). The role of the palm weevil, *Rhynchophorus palmarum*, as a vector of red ring disease of coconuts. I. Results of preliminary investigations. *J. Econ. Entomol.* 56: 375-380.

Hagley E.C. (1964). On the life history and habits of the Palm Weevil, *Rhynchophorus palmarum. Annals of the Ent. Soc. of America* 58 (1): 22-28.

Hagley E.C. (1965). Tests of attractants for the palm weevil. *Journal of Eco. Ent.* 58: 1002-1003.

Hallett R.H., Gries G., Gries R., Borden J.H., Czyzewska E., Oehlschlager A.C., Pierce H.D.Jr., Angerilli N.P.D., Rauf A. (1993a). Aggregation pheromone of two Asian palm weevils, *Rhynchophorus ferrugineus* (Oliv;) and *R. vulneratus* (Panz.). *Naturwissenschaften* 80: 328-331.

Hallet R.H., Oehlschlager A.C., Borden J.H. (1999). Pheromone trapping protocols for the Asian palm weevil, *Rhynchophorus ferrugineus* (Coleoptera: Curculionidae). *Int. J. Pest Manag.* 45 (3): 231-237.

Hallett R.H., Oehlschlager A.C., Gries G., Angerilli N.D.P., Al Shareqi H., Gasouma M.S., Borden J.H. (1993b). Field testing of aggregation pheromone of two Asian palm weevils. Proceedings of the International Conference on oil palm (PORIM), Kuala Lumpur, Malaysia: 661-668

Hardon J.J. (1973). Assisted pollination in the oil palm: a review. In Advances in oil palm cultivation, ISP. Kuala Lumpur.

Henry P. (1948). Les facteurs de la pollinisation chez le Palmier à huile. *Oléagineux* 3 (12): 587-589.

Herms W.B. (1926). *Diocalandra taitensis* of Fanning and Washington islands. *Philippines. J. Sci.* 30 (2): 243.

I.C.A.R (Indian Council of Agricultural Research) (1992). Annual report 1990-1991 p. 71.

Jaffé K., Sanchez P., Cerda H., Hernandez J.V., Jaffé R., Urdaneta N., Guerra G., Martinez R., Miras B. (1993). Chemical ecology of the palm weevil *Rhynchophorus palmarum* (L.) (Coleoptera: Curculionidae): attraction to host plants and to a male-produced aggregation pheromone. *J. Chem. Ecol.* 19: 1703-1720.

Jaya S., Suresh T., Sobhitha-Rani R.S., Sreekumar S. (2000). Evidence of seven larval instars in the red palm weevil *Rhynchophorus ferrugineus* Oliv. reared on sugarcane. *J. Ent. Res.* 24 (1): 27-31.

Jerath M.L. (1965). A list of insects found on palms in Nigeria and their known parasites and predators. *Journal of Nifor* 1969: 411-415.

Kang S.M., A. Karim Z. (1982). Quarantine aspects of the introduction into Malaysia of an oil palm insect pollinator. Proc. Int. Conf. Pl. Prot. in Tropic. Kuala Lumpur: 615-626.

Kloft W.J., Kloft E.S., Kanagaratnam P., Pinto J.L.J.G. (1986). Studies on the use of radioisotopes for the control of the red palm weevil, *Rhynchophorus ferrugineus* F., by the sterile insect technique. *Cocos* 4: 11-18.

Kurian D. Abraham V.A., Ponnamma K.N., Abdullakoya K.M. (1978). *Myllocerus curvicornis* (F). A new pest of coconut. *J. Plant Crops* 6 (1): 44-48.

Lepesme P. (1947). Les insectes des palmiers. Paul Lechevalier Ed. Paris, 903 p.

Lever R.J.A.W. (1937). Economic insects and biological control in the british Solomon islands. *Bull. Ent. Res.* 28: 325.

Lever R.J.A.W. (1941) The small coconut weevil (*Diocalandra*) *Agric. J. Fidji* 12 (4): 3-4.

Lever R.J.A.W. (1969). Les ravageurs du cocotier. *Etudes agricoles de la FAO* n° 77, 196 p.

Louis H., Chandarasekaran N.R. (1980). A note on the weevil *Diocalandra stigmaticollis* as a pest of coconut palm in Tamil Nadu. *Indian Coconut J.* 12: 5-6.

Maharaj S. (1962). Field studies on the life history of the palm weevil. *Journal of the Agricultural Society of Trinidad and Tobago.* June 1962.

Maharaj S. (1973). A new design of trap for collecting the palm weevil, *Rhynchophorus palmarum* (L.) *Ceylon Coconut Plrs. Rev.* 7: 5-7.

Mahindapala R., Subasinghe S.M.P. (1979). Damage to coconut by *Meredolus* Col. Curculionidae. *Ceylon Coconut Quarterly.* 30: 119-120.

Mariau D. (1968). Méthodes de lutte contre le Rhynchophore. *Oléagineux.* 23 (7): 443-446.

Mariau D. (1974). Les ravageurs et maladies du palmier à huile et du cocotier. Le Temnoschoïte. *Oléagineux* 29 (12): 557.

Mariau D. (1985). Rapport de mission entomologique au Brésil. IRHO Doc. 1996 (Document interne).

Mariau D., Desmier de Chenon R. (1991). Les ravageurs du palmier à huile et leurs ennemis en Asie du Sud-Est. *Oléagineux* 46 (11): 400-476.

Mariau D., Desmier de Chenon R., Julia J.F. (1981). Les ravageurs du palmier à huile et du cocotier en Afrique de l'Ouest. *Oléagineux* 36 (4): 168-228.

Mariau D., Genty P. (1988). Contribution de l'IRHO à l'étude des insectes pollinisateurs du palmier à huile en Afrique, en Amérique du Sud et Indonésie. *Oléagineux* 43(6): 233-240.

Mariau D., Houssou M., Lecoustre R., Ndigui B. (1991). Insectes pollinisateurs du palmier à huile et taux de nouaison en Afrique de l'Ouest. *Oléagineux* 46 (2): 43-51.

Meksongsee B., Sakulpanich U. (1975). Field study of trapping coconut weevil *Rhynchophorus vulneratus* (Panzer) with sex attractive. Fourth session of the FAO Technical working Party on Coconut Production and Processing. Kingston Jamaïca. 14-25/9/1979.

Mexzon R.G., Chinchilla C.M., Castrillo G., Salamanca D., (1994). Biologia y habitos de *Rhynchophorus palmarum* L. asociado a la palma aceitera en Costa Rica. *A.S.D. oil palm papers* n° 8.

Mori K., Hiromasa K., Rochat D. (1993a). Synthesis of the stereoisomers of 3-methyl-4-octanol to determine the absolute configuration of the naturally occurring (3S, 4S)-isomer isolated as the male-produced aggregation pheromone of the African palm weevil, *Rhynchophorus phoenicis. Leibigs Ann. Chem.* 1993: 865-870.

Mori K., Kiyota H., Malosse C., Rochat D. (1993b). Synthesis of the enantiomers of syn-4-methyl-5-nonanol to determine the absolute configuration of the naturally occurring (4S, 5S)-isomer isolated as the male-produced pheromone compound of *Rhynchophorus vulneratus* and *Metamasius hemipterus. Liebigs Ann. Chem.* 1993: 1201-1204.

Morin J.P.,Lucchini F., Aranjo J.C.A. de, Ferreira J.M.S., Fraga L.S. (1986). Le contrôle de *Rhynchophorus palmarum* par piégeage à l'aide de morceaux de palmier. *Oléagineux* 41 (2): 57-62.

Moura J.I.L., Mariau D., Delabie J.H.C. (1993). Efficacité de *Paratherezia menezesi* Townsend (Diptera: Tachinidae) pour le contrôle biologique naturel de *Rhynchophorus palmarum* (L.) (Coleoptera: Curculionidae). *Oléagineux* 48 (5): 219-223.

Moura J.I.L., Mariau D., Delabie J.H.C. (1994). Stratégie de lutte contre *Amerrhinus ynca* Shalb 1823 (Coleoptera: Curculionidae), foreur du rachis foliaire du cocotier (*Cocos nucifera* L.) au Brésil. *Oléagineux* 49 (5): 221-226.

Nagnan P., Cain A.H., Rochat D. (1992). Extraction and identification of volatile compounds of fermented oil palm sap (palm wine), candidate attractants for the Palm weevil. *Oléagineux* 47 (3): 135-142.

Nascimento I. A. (1973). Variation in the oxygen consumption of a coconut pest *Homalinotus coriaceus. Boletim de Zoologia et Biologica Marinha (Nova Serie)* 30: 161-183.

Nascimento I. A., De Carvalho H. A. (1970). Morpho physiologic study of the reproductive apparatus of *Homalinotus coriaceus.* Proceedings of the 22nd annual meeting of the sociedade brasileira para o progresso da ciencia, 441 p.

Nirula K.K. (1956). Investigations on the pests of coconut palm part IV. *Rhynchophorus ferrugineus* F. *Ind. Coconut J.* 9: 229-247.

Oehlschlager A.C., Chinchilla C.M., Gonzales LM. (1993c). Optimization of a pheromone-based trap for the American palm weevil, *Rhynchophorus palmarum,* in Proceedings, International Oil Palm Congress, 20-25 September 1993. Palm Oil Research Institute of Malaysia, Kuala Lumpur, Malaysia.

Oehlschlager A.C., Chinchilla C.M., Gonzales LM., Jiron L.F., Prior B., Gries R., Gries G. (1993a). Development of a pheromone based trapping system for *Rhynchophorus palmarum* (Coleoptera: Curculionidae). *J. Econ. Entomol.* 86: 1381-1392.

Oehlschlager A.C., McDonald R.S., Chinchilla C.M., Patschke S.N. (1995a). Influence of a pheromone-based trapping system on the distribution of *Rhynchophorus palmarum* (Coleoptera Curculionidae) in oil palm. *Environ. Entomol.* 24: 1005-1012.

Oehlschlafer A.C., Perez A.L., Pierce H.D.Jr., Gonzales LM., Prior B., Gries R., Gries G. (1993b). Aggregation pheromone of the Asian palm weevil, *Rhynchophorus bilineatus. Naturwissenschaften* 80.

Oehlschlager A.C., Pierce H.D.Jr., Morgan B., Wimalaratne P.D.C., Slessor K.N., King G.G.S., Gries G., Gries R., Borden J.H., Jiron L.F., Chinchilla C.M. (1992). Chirality and field activity of rhynchophorol, the aggregation pheromone of the American palm weevil. *Naturwissenschaften* 79: 134-135.

Oehlschlager A.C., Prior R.N.B.,Perez A.L., Gries R., Gries G., Pierce H.D. Jr., Laup S. (1995b). Structure, chirality, and field testing of a male-produced aggregation pheromone of Asian palm weevil *Rhynchophorus bilineatus* (Montr.) (Coleoptera: Curculionidae) *J. Chem. Ecol.* 21: 1619-1629.

Parasram S., Mederick F. (1971). A note on damage to coconut in St. Lucia, West Indies, by a beetle of the *Cholus zonatus* complex. *Trop. Agric. Trinidad* 48 (2): 125-126.

Perez A.L, Campos Y., Chinchilla C.M., Oehlschlager A.C., Gries G., Gries R., Giblin-Davis R.M.,Castrillo G., Peña J.E, Duncan R.E., Gonzalez L.M., Pierce H.D., McDonald R., Andrade R. (1997). Aggregation pheromones and host kairomones of the West Indian sugarcane weevil, *Metamasius hemipterus sericeus* (Oliv.) (Coleoptera: Curculionidae). *J. Chem. Ecol.* 23: 869-888.

Perez A.K, Gries G., Gries R., Giblin-Davis R.M., Oehlschlager A.C. (1994). Pheromone chirality of the African palm weevil, *Rhynchophorus phoenicis* (F.) and the palmetto weevil, *Rhynchophorus cruentatus* (F.) (Coleoptera: Curculionidae) *J. Chem. Ecol.* 20: 2653-2671.

Perez A.K, Hallett R., Gries R., Gries G., Oehlschlafer A.C., Borden J.H. (1996). Pheromone chirality of the Asian palm weevils, *Rhynchophorus ferrugineus* (Oliv.) and *Rhynchophorus vulneratus* (Panz.), (Coleoptera: Curculionidae). *J. Chem. Ecol.* 22: 357-368.

Philippe R. (1993). Etude de l'incidence des ravageurs sur les inflorescences femelles du palmier à huile en Afrique de l'Ouest. *Oléagineux* 48 (10): 389-405.

Pillai D.M. (1984). Bibliography on red palm weevil (*Rhynchoporus ferrugineus* Fabr.). A pest of coconut (*Cocos nucifera* L.). *Indian Coconut J.* 14 (10): 5-8.

Ponnamma K.N., Kamalakshyamma P.G., Kurian C., Nambiar P.T.N. (1982). Incidence of *Myllocerus curvicornis* (F.) in relation to nutritional status of coconut palms. *PJCS* 7 (1-2): 40-41.

Purohit M.F., Patel G.M., Shah A.H. (1988). First incidence of coconut leaf weevil *Diocalandra frumenti* Fabricius in Gujarat. *Indian Coconut J.* n° 5.

Rahalkar G.W., Harwalkar M.R., Rananavare H.D. (1972). Development of red palm weevil, *Rhynchophorus ferrugineus* Oliv. on sugarcane. *Ind. J. Ent.* 34 (3): 213-215.

Rahalkar G.W., Tamhankar A.J., Shantaram K. (1978). An artificial diet for rearing red palm weevil, *Rhynchophorus ferrugineus* Oliv. *Journal of Plantation Crops* 6 (2): 61-64.

Ramirez-Lucas P., Malosse C., Ducrot P.H., Lettere M., Zagatti P. (1996a). Chemical identification, Electrophysiological and Behavioral Activities of the Pheromone of the *Metamasius hemipterus* (Coleoptera: Curculionidae) *Bioorganic & Medicinal Chemistry* 4 (3): 323-330.

Ramirez-Lucas P., Rochat D., Zagatti P. (1996b). Field trapping of *Metamasius hemipterus* with synthetic pheromone. *Entomol. Exp. Appl.* 80: 453-460.

Rananavare H.D., Shantaram K., Harwalkar M.R., Rahalkar G.W. (1975). Method for the laboratory rearing of red palm weevil. *Rhynchophorus ferrugineus* Oliv. *Journal of Plantation Crops* 3 (2): 65-67.

Restrepo L.G., Rivera F.A., Raigosa B.J. (1982). Ciclo de vida, habitos y morfometria de *Metamasius hemipterus* Oliver y *Rhynchophorus palmarum* L. (Coleoptera: Curculionidae) en caña de azucar (*Saccharum officinarum* L.). *Acta Agron.* 32 (1-4): 33-44.

Risbec J. (1936). les parasites du cocotier en Nouvelle-Calédonie. *Agron. Colon.:* 221-222.

Rochat D. (1990). *Rhyncophorus palmarum.* (Coleoptera: Curculionidae): nuevos datos sobre el comportamiento del insecto y su control par trampeo olfactivo. Perspectivas. *Palmas* 11 (1): 69-79.

Rochat D. (1991). Ecologie chimique du charançon des palmiers, *Rhynchophorus palmarum* L. (Coleoptera: Curculionidae). Thèse de Doctorat de l'Université de Paris VI, 135 p.

Rochat D., Akamou F., Sangare A., Mariau D., Mori K. (1995). Piégeage de *Rhynchophorus phoenicis* (Coleoptera: Curculionidae) à l'aide de Stéréo-isomères de la phéromone d'aggrégation. *C.R. Acad. Sci. Paris.* 318: 183-190.

Rochat D. Descoins C., Malosse C., Nagnan P., Zagatti P., Akamou F., Mariau D. (1993a). Ecologie chimique des charançons des palmiers, *Rhyncophorus spp.* (Coleoptera). *Oléagineux* 48 (5): 225-236.

Rochat D., Gonzalez A.V., Mariau D., Villanueva A.G., Zagatti P. (1991a). Evidence for male-produced aggregation pheromone in American palm weevil, *Rhyncophorus palmarum* L. *Journal of Chemical Ecology* 17 (6): 1221-1230.

Rochat D., Malosse C., Lettere M., Ducrot P.H., Zagatti P., Renou M., Descoins C., (1991b) Male produced aggregation pheromone of the American palm weevil, *Rhyncophorus palmarum* (L.) (Coleoptera: Curculionidae): Collection, Identification, Electrophysiogical activity and Laboratory bioassay. *J.Chem.Ecol.*17 (11): 2127-2140.

Rochat D., Malosse C., Lettere M., Ramirez-Lucas P., Zagatti P. (1993b). Identification of new pheromone-related compounds from volatiles produced by males of four *Rhynchophorinae weevils* (Coleoptera: Curculionidae). *C.R. Acad. Sci. Paris.* Ser.II. 316: 1737-1742.

Rochat D., Nagnan-Le Meillour P., Esteban Duran J., Malosse C., Perthuis B., Morin J.P., (2000). Identification of pheromone synergists in American palm weevil, *Rhyncophorus palmarum*, and attraction of related *Dynamis borassi* (Coleoptera: Curculionidae). *J. Chem. Ecol.* 26 (1): 155-187.

Sadakathulla S. (1991). Management of red palm weevil, Rhyncophorus ferrugineus F. in coconut plantation. *Planter* 67: 415-419.

Simmonds H.W. (1938). Coconut pests and diseases in Melanesia and southern Polynesia. *Depart. of Agric. Fiji Bull.* n° 20.

Sivapragasam A., Arikiah A., Ranjit C.A. (1990). The red stripe weevil, *Rhyncophorus schach* Oliver (Coleoptera: Curculionidae): an increasing menace to coconut palm in Hilir Perak. *Planter* 66: 113-123.

Subramaniam T.R., Thangavel P., David B.V. (1975). A new species of *Amorphoidea* (Curculionidae: Coleoptera) infesting coconut inflorescence. *Curr. Sci.* 44: 825-826.

Syed R.A. (1979). Studies on oil palm pollination by insectes. *Bull. Ent. Res.* 69: 213-224.

Syed R.A. (1982). Insect pollination of oil palm: introduction, establishment and pollinating efficiency of *Elaeidobius kamerunicus* in Malaysia. *Planter* 58: 546-561.

Thajudin S. (2000). Annotated bibliography on *Rhynchophorus ferrugineus* a pest of coconut. *Ind. Coconut Journal* 30 (10): 4-18.

Thompson R.T. (1992). Observations on the morphology and classification of weevils (Coleoptera: Curculionidae) with a key to major groups. *J. Nat. Hist.* 26: 835-891.

Toquica L.S.M. (1993). Insectos diseminadores de *Rhadinaphelenchus cocophilus* (Cobb) Goodey (Tylenchida: Aphelenchoididae) agente causal del anillo rojo-hoja corta en palma de aceite en San Carlos de Guaroa-meta. M.Sc. dissertation. Universidad Nacional de Colombia, 68 p.

Vaurie P. (1966). A revision of the Neotropical genus *Metamasius* (Coleoptera: Curculionidae, Rhynchophorinae). Species groups I and II. *Bull. American Mus. Nat. Hist.* 131: 213-337.

Vaurie P. (1970). Weevils of the tribe Sipalini (Coleoptera, Curculionidae, Rhynchophorinae) Part1. The genera *Rhinostomus* and *Yuccaborus. Am. Mus. Novitates* n° 2419: 1-57.

Vaurie P. (1973). The weevil genera *Homalinotus* and *Ozopherus* of the neotropical cholinae (Coleoptera Curculionidae. *Bull. American Mus. Nat. Hist.* 152: 1-50.

Wattanapongsiri A. (1966). A revision of the genera *Rhynchophorus* and *Dynamis* (Coleoptera Curculionidae). *Depart. of Agric. Science Bull. Bangkok.* I (1), 328 p.

Wilson, M.E. (1962). Laboratory studies on the life history of the palm weevil. *J. of the Agric. Soc. of Trinidad and Tobago* June 1962, 9 p.

Wilson M.E. (1963). Investigations into the Development of the palm weevil *Rhyncophorus palmarum* (L.) *Trop. Agric.* 40 (3): 185-196.

Wolcott G.N. (1933). An economic entomology of the West Indies. Porto Rico.

Zagatti D., Rochat D., Berthier A., Nadarajan L. (1993). Elevage permanent du charançon des palmiers *Rhyncophorus palmarum* in the laboratory. *Oléagineux* 48 (5): 213-217.

OTHER COLEOPTERA

LYMEXYLIDAE

This family only includes one coconut pest, in the Seychelles and Northwest Madagascar.

Melittomma insulare Fairmaire (Vesey-Fitzgerald 1941, Frappa 1948, Brown 1954, Simmonds 1956, Nye 1961)

Description

The adult is cinnamon brown and varies in size from 8 to 20 mm. The larva measures 20 mm in the final instar. The prothorax is very well developed and highly sclerotized, as is the anal segment, which terminates in a horseshoe shape.

Biology

The female only lives for a few days and moves little. Around a hundred eggs are laid in a group in cracks at the base of adult coconut stems, or on the collar of younger plants. The development cycle is not precisely known but is around a year. Pupation takes 2 weeks. No natural enemies have been observed.
The larvae penetrate the stem via a gallery that starts off horizontal then heads upwards. More than 200 larvae can be found in the same stem. Attacks result in the formation of a vast cavity, which weakens the palm, causing it to snap at the base.

Control methods

Parasitoids living on similar species have been sought, without success (Simmonds 1956). There is currently only one curative method, which involves eliminating attacked sections followed by treatment. Studies on the chemical ecology of ***M. insulare*** are required with a view to developing preventive control by olfactory trapping.

SCOLYTIDAE

This family includes numerous species which either attack fallen *Elaeis* fruits on the ground, or mine coconut stems, but none can be considered a serious pest. Most of the species have been observed in Africa, where the fauna of palms has been extensively studied.

Cocotrypes

Lepesme (1947) inventoried 11 species belonging to this genus, 4 of which have been observed on fallen *Elaeis* fruits, particularly in Africa:

C. dactyliperda Fabricius, ***C. congonus*** Eggers (Alibert 1946), ***C. nigripes*** Eggers and ***C. perditor*** Blandford.
The species of the genus ***Stephanoderes*** are mostly found on the rachis or inflorescence: ***S. perhispidus*** Eggers and ***S. punctatus*** Eggers, ***Hypothenemus tuberculosus*** Hagedorn and ***Bothryperus psaltes*** Hagedorn show the same behaviour in the Congo.

Xyleborus
The adults are generally black and small in size (2 mm). Both larvae and adults bore into the stems of *Cocos* and sometimes *Elaeis*. These insects are often considered as debilitating pests, though they can sometimes be primary pests (Paine *in* Lever 1969). The different species are often widespread throughout the neotropical zone.
X. perforans (Wollaston) and ***X. confusus*** Eichhoff have notably been observed in the Seychelles on coconut palms attacked by ***Melittomma insulare*** (Brown 1954). In Brazil, ***X. ferrugineus*** (Fabricius) and ***X. affinis*** Eichhoff indeed seem to be primary pests on coconut. Attacks lead to localized rot and oozing sap (Ferreira and Morin 1985a). We would also mention ***X. torquatus*** Eichhoff and **X. *similis*** (Mahindapala and Subasinghe 1976).

PLATYPODIDAE

Platypus rugulosus Chapuis has been found on coconut palms affected by *stem bleeding* in Puerto Rico (Maramorosh *et al.* 1972).

BUPRESTIDAE

Taphrocerus cocois Bondar (Bondar 1940).
Common in Brazil on coconut palms. The larva mines a narrow elongated gallery in the thickness of the frond; the adult is only 4.5 mm wide. The development cycle lasts 49 days on average. The low economic importance of the insect is probably linked to the existence of two active parasites: *Horismenus* sp. (Hy. *Eulophidae*), on larvae, and *Spilochalcis* sp. (Hy. *Chalcididae*) on pupae (Ferreira and Morin 1985b). In Côte d'Ivoire, it is not rare to see a species of *Buprestidae* on *Elaeis* (Mariau, unpublished). Lastly, we would mention ***Aphamisticus altus*** in Indonesia (Johnston 1965).

CERAMBYCIDAE

In the larval state, numerous species belonging to this family can be found in more or less decomposed *Elaeis* and *Cocos* stems (Lepesme 1947). As these species are not pests, they will not be covered here, though we would mention two species:

Mulciber linnaei is reported in Papua New Guinea on mature and immature palms, though when agornomic conditions are poor (OPRA 1988).

Pterolophia apiceplagiata Breuning can carry out its larva cycle on the husk of coconuts in Sumatra, destroying the sprout if it has already emerged. The development cycle lasts 2.5 to 3 months. Such attacks have proved to be harmful in seed gardens (Desmier de Chenon *et al.* 1991).

NITIDULIDAE

Several species are associated with the male flowers of various palms, two of which are found on *Elaeis*.

Mystrops costaricensis Gillogly. The species has been divided into three sub-species ***Mystrops costaricensis pacificus***, ***Mystrops costaricensis orientalis*** and ***M.c. costaricensis*** depending on their geographical distribution in Latin America (Genty *et al.* 1986). Prior to the introduction of *Elaeidobius kamerunicus* in Latin America, this species ensured excellent pollination on oil palm, which is primarily entomophilous (see *Curculionidae Erirhinae)*. However, the species was nonexistent in several regions in which the original host plants of ***Mystrops*** do not grow, such as *Elaeis oleifera*. ***Mystrops costaricensis*** was introduced into Papua New Guinea (OPRA 1986), but the results of the introduction are unknown.

In Africa, several species of *Nitidulidae* are infrequently seen on male *Elaeis* flowers; they are ***Microporum congolense*** Grouvelle, ***M. dispar*** Watehouse and ***Carpophilus nitidus*** Murray (Desmier de Chenon 1981). ***Microporum*** sp. is sometimes very abundant on *Elaeis* in Madagascar, but it is a mediocre pollinator (Mariau and Genty 1988).

Several species of ***Carpophilus*** have been reported on male coconut flowers in the Asia-Pacific: ***C. dimidiatus*** F. (also a copra pest, Rai and Singh 1977), ***C. foveicollis*** Murray, ***C. humeralis*** F. (Corbett *in* Lepesme 1947).

Lastly, two species have been observed on rotting *Elaeis* bunches in the Congo: ***C. hemipterus*** L. and ***Lasiodactylus costipennis*** Boh. (Ghesquiere *in* Lepesme 1947).

In the same superfamily of the *Cucujoidea* but in other families from those of the *Nitidulidae*, we have found the following on *Elaeis* in Colombia: ***Smicrips*** sp. af.***exilis*** Murray (*Cucujidae*) on female flowers and ***Ahasverus*** sp. (*Silvanidae*) on male flowers (Mariau, unpublished). ***Lithargus*** sp. (*Mycetophagidae*) is frequent on male

Elaeis flowers in Africa (Desmier de Chenon 1981) and, lastly, ***Bitoma palmarum*** Bondar (*Colydiidae*) is sometimes found on *Cocos* flowers (Bondar 1940).

ELATERIDAE

Like many species of Coleoptera, larvae and/or adults are frequently seen in different environments: rotting palm wood, inflorescences; some species can be predators, such as ***Alaus lacteus*** F., which attacks *Oryctes* larvae in India (Webster *in* Lepesme 1947).
Severe attacks have been seen recently in Indonesia (Sumatra) on *Elaeis* roots, by the larvae of an undetermined species (Philippe, personal communication).

COLEOPTERA ON STORED FRUITS

The species involved belong to ten or so families, but studies have mostly been made of those affecting copra (Cotterell 1952, Menon and Pandalai 1958, Mathen 1961, Manson 1974, Rai and Singh 1977, Allotey and Kumar 1988, 1989, Kumari *et al.* 1992). The different species bore into the copra or kernel, causing similar damage. The geographical distribution of these species is usually extremely vast.

Necrobia rufipes De Geer *(Cleridae)* "ham beetle" is one of the most frequently studied species (Laborius *et al.* 1980). Its development cycle varies from 69 to 71 days (incubation: 3.5 days, 5 larval instars: 61 to 66 days, pupation: 6 days). A species of mite feeds off the eggs and young larvae, and pathogens have been isolated from dead larvae (Canete and Gapasin 1980). Temperature, relative humidity and diet have an effect on the development cycle, fecundity and adult life span (Ashman 1962).

Corbett *et al.* (1937 *in* Canete and Gapasin 1980) put forward the figure of 26 species of insects associated with copra in Malaysia, the main one being ***Carpophilus dimidiatus*** F. (*Nitidulidae)* "corn sap beetle", with, in the same genus, ***C. hemipterus*** (L.) "dried fruit beetle" to which we would add ***Trogoderma granarium*** Everts *(Dermestidae)* "khapra beetle".

The family of the *Cucujidae* includes several important species, the most frequently mentioned being ***Oryzaephilus mercator*** (Fauv.) "merchant grain beetle", the commonest species in stored palm kernels in Nigeria (Allotey and Kumar 1988). We would also mention ***O. surinamensis*** L. "saw toothed grain beetle" and ***Ahasverus advena*** Walt. "foreign grain beetle" on copra in India (Kumari *et al.*

1992), along with ***Laemophloeus ferrugineus*** (Steph) "rust-red grain beetle".

Tribolium castaneum (Hbst) (*Tenebrionidae*) "red flour beetle" is a frequently mentioned species, along with ***Lasioderma serricorne*** (F.) (*Anobiidae)* and ***Araecerus fasciculatus*** De Geer. (*Anthribidae*) "coffee bean weevil".

In West Africa, oil palm kernels can be attacked by ***Pachymerus cardo*** (Fåhraeus) (*Bruchidae)* (Prevett 1968), which is often confused with other species, such as ***P. lacerdae*** Chevrolat (Alibert 1946, Castel-Branco 1965), but was clearly identified by Prevett (1966). This species can cause substantial damage in terms of seed viability (Rees 1963).

It is not possible to talk about pests in stored foodstuffs, notably copra, without mentioning the main two Lepidoptera involved: ***Cadra (Ephestia) cautella*** Wlk. "almond moth" and ***Corcyra cephalonica*** Staint. (*Pyralidae)* "rice moth" (Menon and Pandalai 1958) (see Lepidoptera section).

OTHER COLEOPTERA REFERENCES

Alibert H. (1946). Etude sur deux insectes parasites des noix de palme en Afrique occidentale. *Agronomie Tropicale* 1 (3-4): 173-176.

Allotey J., Kumar R. (1988). Insect pest spectrum of stored palm kernel and the damage caused by them. *Insect Sci. Applic.* 9 (5): 617-623.

Allotey J., Kumar R. (1989). Management of palm kernel insect pest in transit sheds. *Discovery and Innovation* 1 (1): 84-90.

Ashman F. (1962). Factors affecting the abundance of the copra beetle *Necrobia rufipes* (Deg.) (Col., Cleridae). *Bull. Ent. Res.* 53: 671-680.

Bondar G. (1940). Insetos nocivos e molestias do coqueiro (*Cocos nucifera*) no Brasil. Bahia, *Tipographia naval*, 160 p.

Brown E.S (1954). The biology of the coconut pest *Melittomma insulare* (Col. Lymexylonidae) and its control in the Seychelles. *Bull. Ent. Res.* 45: 1-66.

Canete M.J.P.,Gapasin D.P. (1980). Biology of the copra beetle, *Necrobia rufipes* De Geer. *Philip. J. Cocon. Stud.* 5 (2): 33-38.

Castel-Branco G.J. (1965). *Pachymerus lacerdae* Chevr. specie nova (Ins. Col. Bruchidae) para a fauna da ilha de S. Tomé. *Garcia de Orta Portugal* 13 (2): 179-184.

Cottterell G.S. (1952). The insects associated with export produce in southern Nigeria. *Bull. Ent. Res.* 43 145-152.

Desmier de Chenon R. (1981). New data on the entomophil pollination of oil palm in West Africa. *Conf. Tech. Oil palm.* AP 11 Kuala Lumpur, 45 p.

Desmier de Chenon R., Asjári M., Kitu W.T. (1991). A newly observed important pest on germinating coconuts in Indonesia *Pterolophia apiceplagiata* Breuning, Coleoptera Cerambycidae. *Oléagineux* 46 (8-9): 321-327.

Ferreira J.M.S., Morin J.P. (1985b). Aspectos de biologia e comportamento do minador da folha do coquiero, *Thaphroceus cocois* Bondar 1922 (Coleoptera: Buprestidae) e ocorrência de parasitismo. *Anais de Sociedade Entomológica do Brasil* Ano 14 n 2.

Ferreira J.M.S., Morin J.P. (1985a). Ocorrência de *Xyleborus ferrugineus* (Fabricius 1801) e Xyleborus affinis (Eichhoff 1867) Coleoptera: Scolytidae) sobre coquiero no estado de Sergipe. *Anais de Sociedade Entomológica do Brasil.* Ano 14 n 2.

Frappa C. (1948). Un parasite du cocotier dans la région nord-ouest de Madagascar. *Agronomie Tropicale* 3 (5-6): 274-281.

Genty P., Garzon A., Lucchini F., Delvare G. (1986). Polinización entomofila de la palma africana en Americá tropical. *Oléagineux* 41 (3): 99-112.

Johnston A. (1965). Host list of fungi and insects recorded in the South East Asia and Pacific regions *Cocos nucifera* L. *Coconut Tech. Document* n 16 FAO, 18 p.

Kumari T.N., Mammen K.V., Mohandas N. (1992). Occurrence and nature of damage caused by pests of stored copra in Kerala. *Indian Coconut J.* 22 (1): 7-9.

Laborius A., Burgstaller H., Von Keyserling K.N. (1989). Storage pests in Tonga and Western Samoa. *Alafua Agr. Bulletin* 5 (3): 59-61.

Lepesme P. (1947). Les insectes des palmiers. Paul Lechevalier Ed. Paris, 903 p.

Lever R.J.A.W.(1969). Les ravageurs du cocotier. *Etudes agricoles de la FAO* n° 77, 196 p.

Mahindapala R., Subasinghe S.M.P. (1976). Dommages causés au cocotier par *Xyleborus similis. Bull. phytosan. FAO* 24 (2): 45-47.

Manson P.D. (1974). Report to the Govt. of Gilbert and Ellice Islands Colony on a survey of insect pests of crops. FAO, UNDP Report, 35 p.

Maramorosch T.R. *et al.* (1972). *Platypus rugulosus* (Platypodidae) and *Xyleborus ferrugineus* (Scolytidae) and certain diseases of coconut palms in Puerto Rico. *J. New York Entomol. Soc. USA* 80 (4): 238-240.

Mariau D., Genty P. (1988). Contribution de l'IRHO à l'étude des insectes pollinisateurs du palmier à huile en Afrique, en Amérique du Sud et Indonésie. *Oléagineux* 43(6): 233-240.

Mathen K. (1961). Storage of coconut products. *Warehousing News Letter* 4: 3.

Menon K.P.V., Pandalai K.M. (1958). The coconut palm. A monograph. *Indian Central Coconut Committee*, Ernakulam, 384 p.

Nye I.W. B. (1961). The control of the coconut pest *Melittomma insulare* (Coleoptera, Lymexylonidae) in Seychelles. *Bull. Ent. Res.* 52: 489-499.

OPRA (1986). Sixth annual report of the Papua New Guinea oil Palm Research Association: 69-82.

OPRA (1988). Annual report of the Papua New Guinea Oil Palm Research Association: 168-169.

Prevett P.F. (1966). The identity of the palm kernel borer in Nigeria, with systemic notes on the genus *Pachymerus* Thunberg (Coleoptera, Bruchidae). *Bull. Ent. Res.* 57: 181-192.

Prevett P.F. (1968). The larva of *Pachymerus cardo* (Fahr.) the palm kernel borer (Coleoptera: Bruchidae). *J. Stored Products Res.* 4 (3): 239-248.

Rai B.K., Singh W. (1977). Studies on insect pests of copra. *Bull. Grain Technol. India* 15 (3): 197-205.

Rees A.R. (1963). Some factors affecting the germination of oil palm seeds under natural conditions. *Jl. W. Afr. Inst. Oil Palm Res.* 4 (14): 201-207.

Simmonds F.J. (1956). An investigation of the possibilities of biological control of *Melittomma insulare* Fairm. (Coleoptera Lymexylonidae), a serious pest of coconut in the Seychelles. *Bull. Ent. Res.* 47: 685-702.

Vesey-Fitzgerald D. (1941). *Melittomma insulare* Fair (Col. Lymexylonidae) a serious pest of coconut in the Seychelles. *Bull. Ent. Res.* 31: 383-398.

ORTHOPTERA

The species of this order belong to the two superfamilies of the *Tettigonioidea*, with a single family (*Tettigoniidae* or katydids) and *Acridoidea*, with two families (*Acrididae* and *Pyrgomorphidae* or crickets). They are usually polyphagous, though some of them seem to be typical palm fauna.

TETTIGONIOIDEA

The family of *Tettigoniidae* comprises 3 genera and 24 species, which are morphologically very similar to each other (Willense 1977, 1979).

Description
Depending on the species, the body length varies from 50 to 85 mm. The males are considerably smaller. The antennae of the adult can be more than three times longer than the body. The female has a long oviscapt that is sometimes curved slightly upwards. They are usually bright green, or more or less dark brown.

Biology
Studies have been concentrated on a few species. ***Sexava coriacea*** L. (Froggatt 1935, Benett *et al.* 1986, Mariau *et al.* 1991, O'Connor 1959, Young 1981), ***Segestidea uniformis*** (Willense) (Room *et al.* 1984) and ***Segestes decoratus*** Redtenbacher (Young 1985).
The 6 to 7 nymphal instars last 4.5 to 6.5 months; the female cycle is longer. Eggs are laid in the ground and incubation varies depending on climatic conditions (2.5 months to 3 months on average) (Young 1985).
The main parasitoids are found on eggs: in the genus ***Sexava***, *Leefmansia bicolor* Waterston (Hym. *Encyrtidae*) is the most important species. This species has undergone mass releases as a biological control method. Other egg parasitoids are less important: *Prosapegus atrellus* Dodd (*Scelionidae*), *Doirania leefmansi* Waterston (Hym. *Trichogrammatidae*), *Ootetrastichus dubius* Waterston (*Eulophidae*) along with two *Mymaridae* and another *Scelionidae* (Lepesme 1947, Froggatt and O'Connor 1940, Baloch 1972). Three species have been found on ***S. decoratus***: *Tetrastichus* sp. and *Tetrastichus* sp. nr. *dubius* (Waterston) (*Eulophidae*), along with the *Scelionidae Triteleia atrella* (Dodd). Parasitism rates vary from 8.7 to 27.7% (Young 1987). The last species has been seen in small numbers on ***Segestidea uniformis*** (Room *et al.* 1984).
The larvae and adult insects are parasitized by the Strepsiptera *Myrmecolacidae, Stichotrema dallatorreanum* Hofeneder. Several

studies have been undertaken to use this parasitoid as a biological control method (Young 1987, Kathirithamby *et al.* 1998, Solulu *et al.* 1998). Biological control trials have also been conducted with gregarines (Hosang *et al.* 1988). Lastly, numerous species of predators have been reported, such as the gecko, *Gehyra oceanica*, snakes such as *Stegonotus modestus* (Rooms *et al.* 1984) and birds such as the common crow *Corvus validus* (Benett *et al.* 1986), but the impact of such predators on populations remains unknown.

Damage and control methods
Attacks primarily occur on foliage, which can be completely destroyed in the event of a serious outbreak, leading to substantial yield losses. Treatment is given by stem injections. The best results have been obtained with methamidophos (Zelazny and Hossang 1988). Attacks have also been seen on nuts, which can cause small nuts to fall.

Species (Willemse 1977-1979)
Sexava: S. karnyi Leefmans in the Togian islands, and nearby Sulawesi, ***S. coriacea*** in the central Moluccas and Papua New Giunea (PNG), ***S. nubila*** (Stäl) in the southern Moluccas and PNG.

Segestes: S. vittacipes Stäl, in the Philippines, where several species exist: ***S. punctipes*** Redtenbacher, ***S. fuscus*** Redtenbacher, ***S. beieri*** Kästner. We would also mention ***S. celebensis*** Karny (Sulawesi), ***S. frater*** Hebard (Moluccas), ***Segestes*** sp. (northern Moluccas), ***S. unicolor*** Redtenbacher (Pacific: Caroline islands, Palau), ***S. decoratus*** Redtenbacher (PNG), ***S. stibicki*** Willemse (PNG) ***S. brevipennis*** Willemse (Irian Jaya) and lastly ***S. cornelli*** Willemse, though seen on *Pandanus,* not *Cocos.*

Segestidea: Apart from ***S. punctipennis*** Bolivar, the other 8 species have been seen in different regions of the Philippines: ***S. novaeguineae*** (Brancsik), ***S. rufipalpius*** (C. Willemse), ***S. uniformis*** (C. Willemse), ***S. gracilis*** (C. Willemse) with two sub-species ***S. g. gracilis*** (C. Willemse) and ***S.g. simulatrix*** Willemse, ***S. leefmansi*** (C. Willemse), ***S. acuminata*** (Kästner), ***S. marmorata*** I. Bolivar, with two sub-species ***S.m. marmorata*** I. Bolivar and ***S.m. occidentalis*** Willemse. Lastly, ***S. soror*** Hebard.

ACRIDOIDEA

Most of the species belong to the family of the *Acrididae* and a few to the family of the *Pyrgomorphidae.*

Crickets are essentially polyphagous at the different stages of their development. However, Seshadri (1956) considered that they prefer monocotyledons to dicotyledons. Several species have been reported as pests, usually on coconut but sometimes on oil palm.

In Asia, different species of the genus ***Valanga*** have been reported. The main species is ***V. nigricornis*** (Burmeister), and serious damage has sometimes been observed in both Malaysia (Han and Chew 1978) and Indonesia (Mariau *et al.* 1991). The development cycle is long, varying from 4 to 6 months (incubation: 4 to 5 weeks, 6 larval instars: 2 to 3 months, adult maturation: 1 month). The adult can be attacked by an *Entomophthora* fungus and the eggs by the larvae of a Coleoptera *Meloidae* (Mariau *et al.* 1991). The other species are, ***V. nigricornis sumatrensis*** Uvarov (in Malaysia) and ***V. transiens*** Walker (in Sabah).

Patanga succincta L., (synonymous with ***V. zehntneri*** Zehntner, in Indonesia), is widespread in the Laccadive islands (Menon 1970).

Locusta migratoria manilensis Meyen can cause damage, but only in the gregarious form (Seshadri 1956). The complete cycle lasts 70 to 110 days (incubation: 15 days, 5 to 6 larval instars: 30 to 50 days). (Mariau *et al.* 1991). Attacks have been reported in the Philippines, Indonesia, Malaysia and the Solomon Islands (Kurian *et al.* 1963).

Other species have also been reported in Malaysia and the Philippines: ***Atractomorpha psittacina*** Hann (*Pyrgomorphidae)*, ***Catantops humilis*** Serv. and ***C. splendens*** Thunberg., ***Oxya velox*** F. and ***O. vicina*** Brunner ***Patanga*** (***Cyrtacanthacris***) ***luteicornis*** Serville, ***Trilophidia annulata*** Thunberg. ***Aularches miliaris*** L. (*Pyrgomorphidae*) has been reported in Sri Lanka and India (Menon 1970) (Kurian *et al.* 1963, Lepesme 1947).

In South America, we would mention ***Schistocerca paranensis*** Burm (Lepesme 1947). Bondar (1940) noted that ***S. flavofasciata*** (De Gueer) is common on coconut and also mentioned on coconut two species of the genus ***Tropidacris***: ***T.cristata grandis*** Thumberg and ***T. collaris*** Stoll. We observed a serious attack by the last species in a young oil palm planting in Brazil (Para) and noted the existence of ***T. cristata latreilli*** Perty on *Elaeis* in Venezuela (Mariau, unpublished).

Lastly, in Africa, the most serious species is ***Zonocerus variegatus*** (L.) (*Pyrgomophidae)* (Vuillaume 1953, 1954, Chapman *et al.* 1977, Toye 1971). At the start of their development, the larvae are almost entirely black, with yellow patches. The dorsal section of the thorax in adults is yellow and the rest of the body is mottled with yellow and red. Diet has little effect on the size of the adult (4.5 cm). In particular, in southern Nigeria, there are two generations per year, one in the dry season, the other in the rainy season (Golding 1940 *in* Iheagwam 1981a). Due to diapause, the egg stage lasts approximately 5 to

6 months depending on the season (Iheagwam 1981b) and the larval instars between 3.5 and 4 months. The female lays 3 to 5 egg cases, each containing 30 to 120 eggs, in the ground. Fecundity varies depending on the adult diet (Iheagwam 1979). There are veritable egg-laying zones in which very high densities of egg cases can be found. The laying site can easily be identified by the mucus left on the soil. Destroying the sites may be a worthwhile control method (Page 1978a). The eggs are parasitized by a Diptera *Blaesoxipha filipjevi* (Rohd) (Page 1978b). ***Z. variegatus*** can cause substantial defoliation in young oil palm plantings, as we saw in Ghana in 1983, but the outbreak was checked by an *Entomophthora* fungus (Mariau unpublished). An epidemic of *Entomophthora grylli* Fresenius was seen to occur within a population that was developing on a poplar clone. In the same population, some individuals were parasitized by *Sarcophaga exuberans* pandelle (*Diptera Sarcophagidae) (*Akanbi 1980). The species is highly polyphagous, but some plants, such as cassava, could prove to be toxic (Mc Caffery *et al.* 1978).

Lastly, we would mention two other species seen on *Cocos* in Nigeria, though of no economic importance for palms: ***Oxycatantops spissus spissus*** Walker and ***Cyrtacanthacris aeriginosa*** (Stoll).

GRYLLOIDEA

We would mention two species of the *Gryllidae* family that can be harmful in nurseries: ***Cardiodactylus novae-guineae*** Hann, in Papua New Guinea and the Solomon Islands (Lever 1969) and ***Amphiacusta caraiba*** Sauss. in Puerto Rico (Wolcott 1936 *in* Lepesme 1947).

ORTHOPTERA REFERENCES

Akanbi M.O. (1980). On the biotic factors causing a mass mortality of *Zonocerus variegatus* L. (Orthoptera: Pygomorphidae) in a poplar clonal nursery in Nigeria. *African J. Plant Protect.* 2 (1): 115-120.

Baloch G.M. (1972). Review of the coconut leafminer and treehopper problems in Papua New Guinea, with a report on their incidence on Southern New Britain. *Papua New Guinea Agricultural J.* 23 (1-2): 27-40.

Bennett C.P.A., Hosang M.L.A., Assa B.H. (1986). Observations of pests and diseases of the coconut, *Cocos nucifera* L. in the Northern Islands of Maluku. *Report coconut Research Institute Manado* Indonesia: 13-24.

Bondar G. (1940). Insetos nocivos e molestias do coqueiro (*Cocos nucifera*) no Brazil. Tipografia Naval, Bahia, 160 p.

Chapman R.F., Cook A.G., Mitchell G.A., Page W.W. (1977). Description and morphometrics of the nymphs of *Zonocerus variegatus* (L.) (Orthoptera, Acridoidea). *Bull. Ent. Res.* 67: 427-437.

Froggatt J.L. (1935). The long-horned tree-hopper of coconuts *Sexava* spp. *New Guinea Agric. Gaz.* 1: 16-27.

Froggatt J.P., O'Connor B.A. (1940). Insects association with the coconut palm. *New Guinea Agric. Gaz.* 6 (3): 16-32.

Han K.J., Chew P.S. (1978). Control measures and observations in some *Valanga nigricornis* outbreaks in oil palms in Peninsula Malaysia. *Planter* 54: 682-691.

Hosang M.L.A., Zekazny B., Tumewan F. (1988). Attempts to release gregarines (Protozoa, Gregarinidae) *Sexana nubila* populations of the Talaud islands. Integrated coconut pest control project. Annual report. Coconut Research Institute Manado North Sulawesi Indonesia: 62-68.

Iheagwam E.U. (1979). Host plant effects on fecundity and adult survival of *Zonocerus variegatus* L. (*Orthoptera, Pyrgomorphidae). Rev. Zool. Afr.* 93 (3): 760-765.

Iheagwam E.U.(1981a). Further evidence for overlap of adults of the dry-and wet-season generation of *Zonocerus variegatus* (*Orthoptera, Pyrgomorphidae*) from durations of nymphal development. *Rev. Zool. Afr.* 95 (2): 458-464.

Iheagwam E.U.(1981b). On some aspects of the biology of immature stages of the grasshopper pest, *Zonocerus variegatus* L. (Orthopt., Pyrgomorphidae). *Sonderbruck aus Bd.* H. 2, S.: 149-154.

Kathirithamby J., Simpson S., Solulu T., Caudwell R. (1998). Strepsiptera parasites-novel biocontrol tools for oil palm integrated management in Papua New Guinea. *International Journal of Pest Management* 44 (3): 127-133.

Kurian C., Pillai G.B., Menon K.S. (1963). Locusts a problem to coconut. *Indian Coconut J.* 16 (3): 127-136.

Lepesme P. (1947). Les insectes des palmiers. Paul Le Chevalier ed. Paris, 903 p.

Lever R.J.A.W. (1969). Les ravageurs du cocotier. *Etudes agricoles de la FAO* n° 77, 196 p.

Mariau D., Desmier de Chenon R., Sudharto P.S. (1991). Les ravageurs du palmier à huile et leurs ennemis naturels en Asie du sud-est. *Oléagineux* 46 (11): 400-476.

Mc Caffery A.R., Cook A.G., Page W.W., Perfect T.J. (1978). Utilisation of food by *Zonocerus variegatus* (L.) (Orthoptera: Acridoidea). *Bull. Ent. Res.* 68 (4): 589-606.

Menon K.S. (1970). Bombay locust a pest from the Laccadives. *Coconut Bull. India* 1 (1): 8-9.

O'Connor B.A. (1959). The coconut treehopper, *Sexava* sp. and its parasits in the Madang district. *Papua New Guinea Agricultural J.* 11 (4): 121-125.

Page W.W. (1978a). Destruction of eggs as an alternation to chemical control of the grasshopper pest *Zonocerus variegatus* (L.)(Orthoptera, Acridoidea). *Bull. Ent. Res.* 68 (4): 575-581.

Page W.W. (1978b). The biology and control of the grasshopper *Zonocerus variegatus. Pans* 24 (3): 270-277.

Room P.M., Perry C.M., Bailey P.P. (1984). A population study of the coconut pest *Segestidea uniformis* (Willemse) (Orthoptera: Tettigonioidea) on an equatorial island. *Bull. Ent. Res.* 74: 439-451.

Seshadri A.R. (1956). An outbreak of the migratory locust (*Locusta migratoria*) in Ramanathapuram District (Madras) in 1954. *Indian J. Ent.* 18: 112-122.

Solulu T.M., Simpson S.J., Kathirithamby J. (1998). The effect of strepsipteran parasitism on a tettigoniid pest of oil palm in Papua New Guinea. *Physiological Entomology* 23: 388-398.

Toye S.A. (1971). Notes on the biology of *Zonocerus variegatus* (L.) (Orthoptera, Acridoidea) in the Western State of Nigeria. *Revue Zoologie Africaine* 86: 384-392.

Vuillaume M. (1953). Biologie et comportement en AOF de *Zonocerus variegatus.* Thesis University of Paris.

Vuillaume M. (1954). Etude du cycle de *Zonocerus variegatus* en basse Côte d'Ivoire. *Fruits d'Outre Mer* 9: 147-156.

Willemse F. (1977). Classification and distribution of the *Sexavae* of the melanesian subregion (Orthoptera, Tettigonioidea, Mecopodinae) *Tijdschrift voor Entomologie* 120 (8): 213-277.

Willemse F. (1979). Additional notes on the *Sexavae* of the melanesian subregion (Orthoptera, Tettigonioidea, Mecopodinae). *Entomologische Berichten* 39 (1): 4-9.

Young G.R. (1981). *Sexava*: a pest of coconut and oil palm *Papua New Guinea Dep. Primary Ind. Information Bull.* n 37, 3 p.

Young G.R. (1985). Observations on the biology of *Segestes decoratus* Redtenbacher (Orthoptera: Tettigoniidae) a pest of coconut in Papua New Guinea. *Gen. Appl. Ent.* 17: 57-64.

Young G.R. (1987). Some parasites of *Segestes decoratus* Redtenbacher (Orthoptera: Tettigoniidae) and their possible use in the biological control of Tettigoniid pests of coconut in Papua New Guinea. *Bull. Ent. Res.* 77: 515-524.

Zelazny B., Hosang M.L.A. (1988). Ecological studies on *Sexava* spp. and discussion on control with pecticides. Integrated coconut pest control Project. Annual Report. Coconut Research Institute Manado, North Sulawesi, Indonesia: 69-78.

PHASMIDA

This order only has one family *(Phasmidae),* which includes some species attacking palms.

Graeffea crouanii Le Guillou (***G. cocophaga*** Newport) (Rapp 1989, Bedford 1978, Swaine 1969, Paine 1968, Lever 1944, O'Connor *et al.* 1954, Singh 1981).
This is the most important species and the most widespread on numerous Pacific islands: Fiji, Tonga, Caroline, Marquesas, Vanuatu, New Caledonia, etc ***G. crouanii*** is polyphagous but its main host is coconut.

Description
Both sexes are greenish-brown and the female (110 mm long) is much larger than the male (67 mm). All the species in this family have a very elongated shape. The overall appearance is that of a stick. Their wings are very little developed and the female is flightless.

Biology
The development cycle lasts 7.5 months (O'Connor *et al.* 1954), 95 days for incubation, 108 days for the 6 female larval instars and 94 days for the 5 male instars. The pre-oviposition period lasts 20 days. The eggs are sensitive to temperatures between 20 and 30°C and hatching rates are higher in very weedy plantations. Fecundity can be very high (up to about 300 eggs). Most of the eggs laid fall to the ground, and the young larvae then climb up the stem to the leaf crown. Other eggs incubate at the base of fronds.

Natural enemies
Parasitoids
Two species of Hymenoptera *Eupelmidae, Paranastatus verticalis* Eady and *P. nigriscutellatus* Eady, which parasitize 10 to 25% of eggs (Paine 1968), are indigenous to Fiji (Eady 1956). These two parasitoids have been introduced on several Pacific islands, sometimes successfully, as in Tonga (Crooker 1979 *in* Rapp 1989). A Diptera *Sarcophagidae* has been obtained from an adult in Fiji (Taveuni) (Singh *et al.* 1973 *in* Rapp 1989). *Drino* sp. (Dipt. *Tachinidae*) can parasitize up to 11% of phasmids in Fiji (Singh 1972 *in:* Rapp 1989).

Predators
Numerous types of predators attack phasmid eggs: lizards, birds, spiders.
(*Ascyltus pterygotes* Koch, *Salticidae*) (Rapp. 1989). Several species of ants can destroy up to and over 50% of eggs: *Pheidole*

megacephala Fabricius, *Odontomachus simillimus* Fr., *Monomorium floricola* Jerdon, *Solenopsis geminata* Fabricius, *Tapinoma melanocephalum* Fabricius, *Anoplolepis longipes* Jerdon, etc (Taylor *in:* Rapp. 1989).

Control methods (O'Connor 1959).
When natural enemies are not enough to check populations, chemical treatment becomes necessary. Injection with monocrotophos or dicrotophos has given excellent results (Stelzer 1970, Darmaradjan 1977). Residues remaining in coconut water are below the accepted level (Patel *et al.* 1985). Results in trials with other systemic insecticides have not been so good (Rapp. 1989).

Other species
The biology of these species is very similar to that of ***G. crouanii*** but they are spatially much more localized: ***G. seychellensis*** Ferrari, in the Seychelles, ***G. lifuensis*** Sharp in the Loyalty Islands (Lever 1969).

Ophicrania leveri Günther develops in the Solomon Islands, where outbreaks only occur on the island of Savol (Lever 1969). Two species of Diptera *Tachinidae* parasitoids have been noted: *Mycteromyiella phasmatophaga* (Crosskey 1966) and *M. laetifica* (Mesnil) (Paine 1968).

Megacrania phelaus Westwood is still limited to the island of Ugi (Solomons) (Lever 1969).

In Oro province (Papua New Guinea), outbreaks of ***Eurycantha insularis*** Lucas have been treated by monocrotophos injection (OPRA 1989).

PHASMIDA REFERENCES

Bedford G.O. (1978). Biology and ecology of the Phasmatodea. *Ann. Rev. Entomol.* 23: 125-149.

Catala R. (1948). Sur une grave menace que fait peser sur les cocoteraies des Nouvelles Hébrides l'extension de *Graeffea cocophaga. New. Rev. Int. Bot. Appl.* 309-320: 354-357.

Crosskey R.W. A new species of *Mycteromyiella* (Diptera: Tachinidae) parasitic on *Ophieramia leveri* (Günther) (Phasmida: Phasmatidae) in the Solomon islands. *Bull. Ent. Res.* 57: 525-532.

Dharmarajan E. (1977). Trunk injections with systemic insecticides for the control of the coconut stick insect *Graeffea crouanii* (Le Guillou) *Alafua Agric. Bull.* 2 (2): 6-7.

Eady R.D. (1956). Two new species of the genus *Paranastatus masi* (Hym. Eupelmidae) from Fiji. *Bull. Ent. Res.* 47: 61-67.

Lever R.J. A.W. (1944). The coconut stick-insect. *Agric. J. Dep. Agri. Fiji* 15: 18-19.

Lever R.J. A.W. (1969). Les ravageurs du cocotier. Etudes agricoles de la FAO n 77, Rome, 196 p.

O'Connor B.A. (1959). La lutte contre le phasme du cocotier aux Fidji. *Bull. Pacifique Sud.* 10 (1): 23-30.

O'Connor B.A., Pillai J.S., Singh S.R. (1954). Notes on the coconut stick insect *Graeffea crouanii* (Le Guillou), *Fiji Agri. J.* 25 (1): 89-92.

OPRA (1989). Annual report of the Papua New Guinea Oil Palm Research association.

Paine R.W. (1968). Investigations for the biological control in Fidji of the coconut stick-insect *Graeffea crouanii* (Le Guillou). *Bull. Ent. Res.* 57: 567-604.

Patel R.K., Shah A.H., Jhala R.C. (1985). Residue status of monocrotophos in coconut water injected by stem injection technique. *GAU Res. J.* (10) 2: 16-22.

Rapp G. von (1989). Coconut stick insect *Graeffea crouanii* (Orthoptera: Phasmida), Coconut flat moth *Agonoxena argaula* (Lepidoptera: Agonoxenidae). Studies on pest status, biology, ecological control carried out in the kingdom of Tonga, South Pacific. *Dissertation zur Erlangung des Grades eines Doktors der Agrar wissenschaften I*. Stuttgart. Hohenheim, 114p.

Singh S.R. (1981). Ecological studies on coconut stick insect *Graeffea crouanii* (Le Guillou) (Phasmatoidea: Phasmatidae). *Ann. Res. Rept. Dep. Agric. Fiji*: 72-74.

Stelzer M.J. (1970). Preliminary studies on the control of the coconut stick insect, *Graeffea crouanii* (Le Guillou), with systemic insectides. *Bull. Ent. Res.* 60: 49-51.

Swaine G. (1969). The coconut stick insect *Graeffea crouanii* (Le Guillou). *Oléagineux* 24 (2): 75-77.

ISOPTERA

Numerous species of termites are found in coconut and oil palm plantations; they usually attack pieces of dead plants and do not therefore have a negative effect on the palms. It has been shown in a coconut plantation in Côte d'Ivoire, that termites and earthworms constitute the greatest biomass of the macrofauna in the soil (Kouassi and Lepage 1988).

However, many species attack living tissues, which has become a fairly general phenomenon in recent years on many other plants, and about which the specialists can only put forward hypotheses.

TERMITIDAE

The family of the *Termitidae* contains the largest number of species. Kashyap *et al.*(1984) mentions 22 on coconut.

In the genus ***Macrotermes***, we would mention ***M. bellicosus*** Smeathman (***Bellicositermes nigeriensis*** Sjostedt), which mines young coconut plants in the nursery (Aisagbonhi 1989) and in the first year after planting (Mariau 1971, Agodan 1980). ***M. subhyalinus*** is found on coconut in Africa (Rambur) (Harris 1969), and ***M. carbonarius*** (Hagen) (Harris 1969) and ***M. gilvus*** Hagen in Malaysia (Lever 1969).

In Côte d'Ivoire, several species can attack young palm plantings, the main one being ***Amitermes evuncifer*** Silvestri. In Malaysia, ***A. globosus*** and ***A. sulphureus*** Hav. are humivorous (Lepesme 1947), but there is also a species of the genus ***Odontotermes*** (Han *et al.* 1998). In the latter genus ***O. obesus*** (Rambur) is the most common species on coconut in India. Nest construction by this species has been precisely studied (Veeranna and Basalingappa 1989). In young plantings, up to 20% attacks have been seen on lateritic soil (Kurian and Mathen 1964, Mohan 1999). Fernando (1962) mentions ***O. obsauriceps*** Wasmann and **O. *redemanni*** Wasmann in Sri Lanka, to which should be added ***O. horni*** Wasmann (Harris 1969). In Tanzania, it is ***Allodontermes morogorensis*** Harris (Harris 1969) that attacks coconut, whilst ***Hypotermes*** sp. and ***Pericapritermes*** sp. attack the roots of *Elaeis guineensis* in Indian nurseries (Dhileepan 1992).

Many species are found in coconut and oil palm plantations, where they sometimes construct very large nests along the stems. However, these species are barely harmful, if at all. Such is the case with ***Microcerotermes biroi*** Desneux, whose population dynamics and reproduction have been studied (Leponce *et al.* 1996). In New

Guinea, this species lives in coconut plantations along with ***Nasutitermes princeps*** and ***N. novarumhebridarum*** Holmgren (Leponce *et al.* 1997). These species have also been reported in the Solomon Islands (Harris and Brown 1960). That is also the case with ***Syntermes*** sp. in Brazil, which even seems to have a beneficial effect on *Elaeis* (Nogueira and Moura 1984).
In the genus ***Microcerotermes***, we would also mention ***M. bugnioni*** Holmgren in Sri Lanka, ***M. paraffinis*** Silvestri in Samoa and ***M. los-bañosensis*** Oshima in the Philippines, ***M. piliceps*** Snyder in the Solomon Islands and ***M. subtilis*** Wasmann in the Seychelles (Lepesme 1947).

Many other species of ***Nasutitermes*** have been reported on coconut: ***N. ceylonicus*** Holmgren in Sri Lanka, ***N. ephratae*** Holmgren in Panama and ***N. laticeps*** Wasmann in Madagascar, which would seem to encourage attacks by borers (Lever 1969). We could also mention ***N. costalis*** Holmgren, which is common in Puerto Rico, ***N. negritus*** Wasmann in the Seychelles, which follows attacks by *Melittomma insulare* Frm (Col. *Lymexylidae*), ***N. lugonicus*** Oshima in the Philippines (Lepesme 1947). The most common species in Brazil is ***N. (Eutermes) ripertii*** (Rambur) (Bondar 1940). On oil palm in Malaysia, Wood (1968) reported ***N. havilandi*** (Desneux) and ***N. matangensiformis*** (Holmgren), which do not attack living tissues either.

In the genus ***Termes***, ***T. classicus*** Sjöst attacks germinated nuts in Somalia (Paoli 1931 *in* Lepesme 1947), whilst ***T. carbonavius*** Hog.has been reported on coconut stems in Malaysia (Dammerman 1929, *in* Lepesme 1947).

RHINOTERMITIDAE

It is in the family of the *Rhinotermitidae* that a major Asian oil palm and coconut pest is found: ***Coptotermes curvignathus*** Holmgren, which leads to the death of numerous palms in Indonesia and Malaysia (Kalshoven 1963). In peat soil, this species has indirectly led to the death of tens of thousands of coconut palms (Mariau *et al.* 1992). The workers climb the stem and penetrate the coconut palm, in its apical section, boring into forming tissue. This provides access to bacteria that kill the palm within a few days. Serious attacks have also been reported on *Elaeis* (Mariau *et al.* 1991).
The other species in the same genus are ***C. ceylonicus*** Holmgren, which attacks coconut stems in Sri Lanka. ***C. travians*** is reported on oil palm in Malaysia (Wood 1968) and ***C. truncatus*** Wasmann attacks coconut palms in the Seychelles after attacks by the Coleoptera *Melittomma* (Lever 1969). Three other species have been observed on *Cocos:* ***C. niger*** (Snyder) in Panama, ***C. heimi*** (Wasmann) in India

(Bose and Das 1982) and ***C. sinabangensis*** Oshima in Indonesia (Harris 1969).

Lastly, three species of the genus ***Schedorhinotermes*** are mentioned with ***S. marjoriae*** Synder (Harris and Brown 1960) on coconut and ***S. longirostris*** Brauer and ***S. malaccensis*** Holmgren on oil palm in Malaysia (Corbett 1932 *in* Lepesme 1947), but no damage has been reported.

KALOTERMITIDAE

This family includes ***Calotermes (Neotermes) rainbowi*** Hill, which is a serious coconut pest on some of the Cook Islands (Hopkins 1927, Kelsey 1945), where virtually all the coconut palms can be affected to some degree (Given 1964). ***C. laticollis*** Holmgren is a secondary attacker on coconut palms affected by *Melittomma* (Vesey-Fitzgerald 1941 *in* Lepesme 1947) and ***C. immigrans*** Snyder in the Christmas archipelago (Witney 1929 *in* Lepesme 1947). ***C. heimi*** (Wasmann) can be serious on coconut in the State of Orissa, India (Bose and Das 1982).

Termite control methods

Termites are highly susceptible to insecticides in the organochlorine and organophosphorous groups (Sadakathulla and Ramachandran 1990, 1991, Han *et al.* 1998, Mariau *et al.* 1992). Logan *et al.* (1990) indicate the potential control methods available: cultural techniques promoting plant development, plant-derived insecticides, use of predators and pathogens. Indeed, numerous termite species attack young plantings in difficult agroclimatic conditions. Termites are susceptible to pathogens and particularly to fungi. For instance, ***Coptotermes curvignathus*** is 100% destroyed by a *Conidiobolus* (Alstom 1947), but in Petri dishes. However, no termite specialist has yet reported any termite nests affected by fungi in the wild (Grassé 1986).

ISOPTERA REFERENCES

Agodan A. (1980). Pests and diseases of the oil palm and coconut. New control methods against termites harmful to the coconut in West Africa. *Oléagineux* 35 (3): 145-148.

Aisagbonhi C.I. (1989). A survey of the destruction effect of *Macrotermes bellicosus* Smethman (Isoptera: Termidae, Macrotermidae) on coconut seednuts at NIFOR, Benin, Nigeria. *Tropical Pest Management* 35 (4): 380-381.

Alstom R.A. (1947). A fungus parasite on *Coptotermes curvignatus* Holm. *Nature* 160: 120.

Bondar G. (1940). Insectos nocivos et molestias do coquiero (*Cocos nucifera*) no Brasil. *Tipografia naval*, Bahia, 160 p.

Bose G., Das B.C. (1982). Termite fauna of Orissa state, Eastern India. *Zoological survey of India* 80 (1-2): 197-213.

Dhileepan K. (1992). Insect pests of oil palm *(Elaeis guineensis)* in India. *Planter* 68: 183-191.

Fernando H.E. (1962). Termites of economic importance in Ceylan. *In*: "Termites in Humid Tropics" (Proc. New Delhi Symp.): 205-210. Paris.

Given R.B. (1964). Le termite du cocotier aux îles Cook. *Bull. Pacifique Sud* 14 (3): 11-12.

Grassé P.P. (1986). Termitologia. Prédateurs et parasites de termites, tome III, Masson ed.: 205.

Han S.H., Tokro G.P.,Tano Y., Lepage M. (1998). Dégâts des termites dans les plantations de palmiers à huile en Côte d'Ivoire : évaluation et méthodes de lutte. *Plantations, recherche, développement* 5 (2): 119-126.

Harris W.V. (1969). Termites as pests of crops and trees. *Commonwealth Insitute of Entomology,* Queen's Gate London, 41 p.

Harris W.V., Brown E.S. (1960). The termites of Solomon Islands. *Bull. Ent. Res.* 49: 737-750.

Hopkins G.H.E. (1927). Pests of economic plants in Samoa and other island groups. *Bull. Ent. Res.* 18: 18-32.

Kalshoven L.G.E. (1963). *Coptotermes curvignathus* causing the death of trees in Indonesia and Malaya. *Entomologishe Beritchen Dell* 231. V.: 90-100.

Kashyap R.K., Verma R.N., Bhanot J.P. (1984). Termites of plantation crops and their damage and control. A review article. *J. Plant. Crops* 12 (1): 1-10.

Kelsey J.M. (1945). A termite damaging coconut palms on Suwarro islands: *Calotermes (Neotermes) rainbowi* Hill. *New Zeeland Journ. Sci. And Tech.* 27 (1): 69-75.

Kouassi P., Le Page M. (1988). Comparative seasonal development of termite populations of Guinean ecosystems (Ivory Coast). Acte des colloques insectes sociaux 4: 330-340.

Kurian C., Mathen K. (1964). Underground saboteurs of coconut. *Cocon. Bull.* 18 (5): 223-230.

Leponce M., Roisin J., Pasteels J.M. (1996). Reproduction mechanisms and dynamics of habitat colonization in *Microcerotermes biroi* (Isoptera: Termitidae). *Ecol. Entomol.* 21: 178-184.

Leponce M., Roisin Y., Pasteels J.M. (1997). Structure and dynamics of the arboreal termite community in New Guinea coconut plantations. *Biotropica* 29 (2): 193-203.

Lepesme P. (1947). Les insectes des palmiers Paul Lechevalier ed., Paris, 903 p.

Lever R.J. A.W. (1969). Les ravageurs du cocotier. Etudes agricoles de la FAO n 77, Rome, 196 p.

Logan J.W.M., Cowie R.H., Wood T.G. (1990). Termite (Isoptera) control in agriculture and forestry by non-chemical methods: a review. *Bull. Ent. Res.* 80 (3): 309-330.

Mariau D. (1971). Les ravageurs et maladies du palmier et du cocotier. Méthodes de lutte contre les termites attaquant les jeunes plants de cocotier. *Oléagineux* 26 (4): 233-234.

Mariau D., Desmier de Chenon R., Sudharto P.S. (1991). Les ravageurs du palmier à huile et leurs ennemis en Asie du Sud-Est. *Oléagineux* 46 (11): 400-476.

Mariau D., Renoux J., Desmier de Chenon R. (1992). *Coptotermes curvignatus* Holmgren Rhinotermitidae principal ravageur du cocotier planté sur tourbe à Sumatra. *Oléagineux* 47 (10): 561-568.

Mohan C. (1999). Know more about the occasional pests of coconut palm. *Indian Coconut J.* 30 (6): 4-9.

Nogueira S.B., Moura J.I.L. (1984). Dendezeiro (*Elaeis guinensis*): en associaçao com o " Capim de terrra solta ", *Syntermes* sp. (Isoptera: Termitidae) un mutualismo interessante. *Floresta* 15 (1-2): 50-51.

Sadakathulla S., Ramachandran T.K. (1990). Evaluation of insecticides against termite *Odontotermes obesus* Ramb (Termitidae: Isoptera) attacking palm and plaited leaves of coconut. *Indian Coconut J.* 21 (1): 11-13.

Sadakathulla S., Ramachandran T.K. (1991). Termites should no longer worry you in coconut plantations. *Indian Horticulture* oct. déc.: 24.

Veeranna G., Basalingappa S. (1989). Nesting pattern of the termites *Odontotermes obesus* Rambur and *Odontotermes wallonensis* Wasmann (Isoptera: Termitidae). *Insect. Sci. Applic.* 10 (2): 169-180.

Wood B.J. (1968). Pests of oil palm in Malaysia and their control. The incorporated society of planters, Kuala Lumpur, 204 p.

THYSANOPTERA

Whilst thrips can be serious pests and disease vectors on numerous plants, they only play a highly secondary role on palms. However, this type of insect is usually quite common on flowers, primarily those of the coconut palm, but some of them are mycophagous. ***Haplothrips ceylonicus*** Schmutz is found in Malaysia (Miller 1932 *in* Lepesme 1947). Also worth mentioning on coconut are ***Hindsiana cocois*** Watson (***Karnyothrips melaleucus*** Bagnall), ***H. ceylonicus*** Schmutz and ***H. arbensis*** Wats in Cuba, ***Diceratothrips bicornis*** Bagnall, ***Eupathithrips silvestrii*** (Buffa) and ***Dichaetothrips brevicollis*** Hood in Guiana. ***Nesothrips brevicollis*** (Bagn) and ***Scotothrips claripennis*** (Moulton) (***Ethirothrips brevis*** (Bagnall)), which are widespread and have been observed on coconut fronds in Florida (Reinert and Nakahara 1976). In India, ***Kleothrips subramani*** (Ramakrishna) (***Tiarothrips subramanii*** (Ramakrishna)) and ***Symphiothrips punctatus*** Hood and Williams have also been seen on coconut fronds (Lepesme 1947). ***Heliothrips haemorrhoidalis*** Bouché, which is very common in greenhouses, is highly polyphagous and widespread, and is reported on coconut fruits in Brazil (Bondar 1940).

Lastly, though this is not an exhaustive list, three species have been observed on coconut inflorescences in Jamaica*:* ***Franklinella bruneri*** Watson, ***F. kelliae*** Sakimura and ***F. insularis*** (Franklin) (Sakimura 1986). In central Africa, several species have been reported on *Elaeis* (Lepesme 1947).

We would also mention ***Thrips hawaiiensis*** (Morgan), which was considered to be the main pollinator of oil palm in the Indo-Malay peninsula (Syed 1981). Since the introduction of the *Curculionidae Elaeidobius kamerunicus* (see Col. *Curculionidae*) its pollinating role has virtually disappered. Its development cycle lasts 9 days and the insect completes the entire cycle on the palm.

THYSANOPTERA REFERENCES

Bondar G. (1940). Insetos nocivos et molestias do coquiero (*Cocos nucifera*) no Brasil. *Tipografia naval*, Bahia, 160 p.

Lepesme P. (1947). Les insectes des palmiers, Paul Lechevalier ed., Paris, 903 p.

Reinert J.A., Nakahara S. (1976). *Nesothrips brevicollis* and *Scotothrips claripennis* (Thysanoptera: Phlaeothripidae) on *Cocos nucifera*: two new species records for the continental United States. *Florida Entomologist* 59 (3): 308.

Sakimura K. (1986). Thrips in and around the coconut plantations in Jamaica, with a few taxonomical notes (Thysanoptera). *Florida Entomologist* 69 (2): 348-363.

Syed R.A. (1981). Pollinating thrips of oil palm in West Malaysia. *Planter* 57-659: 62-81.

HYMENOPTERA

The entomofauna of oil palm and coconut plantations includes a very large number of Hymenoptera parasitoids, which have been studied with their respective hosts, and *Formicidae* predators also mentioned with their prey. However, some ants can have a negative indirect effect, by encouraging the development of certain aphids and scale insects, such as, among others, ***Azteca*** sp. in Guyana (Rai 1977) or more directly ***Anoplolepis longipes*** Jerd., which excavates the soil in India and can kill certain coconut palms (Veeresh and Gubbaiah 1984). The latter species also has a negative effect by eliminating the red ant *Oecophylla,* which is an active predator of the sucking Hemiptera *Amblypelta cocophaga* China. We would also mention ***Trigonata hyalinata*** var. ***branneri*** Cockerell (*Apidae*), which attacks male and female coconut flowers in Suriname (Dinther 1956). Other species can be very aggressive towards man and considerably disrupt agricultural work, such as ***Nylanderia fulva*** (Mayr) in Colombia (Zenner de Polonia and Bolaños 1982).

A single species is a primary pest on palms, especially on *Elaeis*: ***Atta cephalotes*** L. It is common in the oil palm plantations of Latin America. The workers cut out pieces of leaflets to make a mushroom bed used to cultivate a fungus belonging to the genus *Leucocoprinus* on which they feed. This ant can cause substantial defoliation (Genty *et al.* 1978, Cardenas 1992, Zenner de Polonia 1994). It is mostly the fronds in the upper crown that are attacked. This is to do with the composition of the fronds, as the older ones are richer in lipids (Littledyke and Cherrett 1978).

HYMENOPTERA REFERENCES

Cardenas M.R. (1992). Manejo de la hormiga arriera *Atta cephalotes* (L.) en zonas cafeteras de Colombia. *In:* Hormigas: Caracteriticas, danos y manejo, *Sociedad Colombiana de Entomologia, Miscelanea (Colombia)* 24: 23-31.

Dinther J.B.M. van (1956). Insects of coconut palm in Suriname. *Landbouwproefstation in Suriname.* Bulletin N° 69, 27 p.

Genty Ph., Desmier de Chenon R., Morin J.P. (1978). Les ravageurs du palmier à huile en Amérique latine. *Oléagineux* 33 (7): 325-419.

Littledyke M., Cherrett J.M. (1978). Defence mechanismes in young and old leaves against cutting by the leaf cutting ants *Atta cephalotes* (L.) and *Acromyrmex octopinosus* (Reich) (Hymenoptera: Formicidae). *Bull. Ent. Res.* 68: 263-271.

Rai B.K. (1977). Damage to coconut palms by *Azteca* sp. (Hymenoptera: Formicidae) and insecticidal control with bait in Guyana. *Bull. Ent. Res.* 67: 175-183.

Veeresch G.H., Gubbaiah. (1984). A report on the “crazy ant” (*Anoplolepis longipes*) menace in Karnataka. *Journal of Soil Biology and Ecology* 4 (1): 65-73.

Zenner de Polonia I. (1994). Hormigas depredadores en le ecosistema de palma de aceite. *Palmas* 15 (4): 33-39.

Zenner de Polonia I., Bolaños N.R. (1982). Uso de cebos contra la *hormiga loca Nylanderia fulva* (Mayr) (Hymenoptera: Formicidae). *Revista Colombiana de Entomologia* 8 (1-2): 24-31.

MITES

Virtually all mite pests belong to the two families of the *Tetranychidae* and the *Eriophyidae*. Many other mites have been reported on palms, but they are either predators or feed off debris (Ghai and Wadhi 1983, Mariau 1989, Sathiamma 1996).

TETRANYCHIDAE

The mites in this family, which are often referred to as red spiders, puncture the plant in their larval and adult forms. They are usually more or less bright red. Attacks mostly occur on the foliage of young palms in nurseries, or the early years after planting. Individuals often weave a silken shelter. An inventory of 8 species was drawn up on coconut (Ghai and Wadhi 1983). Several other species have been reported, with two main genera: ***Oligonychus*** and ***Tetranychus***. In India, it is the species ***O. iseilemae*** (Hirst) and ***T. ludeni***, Zacher that are most frequently mentioned (Sathiamma 1996), but also ***T. hindustanicus*** Hirst (CPCRI 1979) and perhaps ***Schizotetranychus hindustanichus*** (Hirst) (Ghai and Wadhi 1983).
O. iseilemae (Hirst) is highly polyphagous, but coconut is its only perennial host. The length of the development cycle, the life span of the females and fecundity vary considerably depending on the host plant (Sathiamma 1988).
T. ludeni has a development cycle of 8 to 11 days and is most frequently found in January-February (Sathiamma 1988). ***O. indicus*** (Hirst) has also been seen on coconut fronds in India (Channa Bassavana and Banu 1972), along with ***T. fijiensis*** Hirst, though it is widely distributed since this mite was reported for the first time in Fiji (Hirst 1924 *in* Ghai and Wadhi 1983), then in Thailand (Baker 1976), the Seychelles (Gutierez 1974), Sri Lanka (Kanagaratnam *et al.* 1979 *in* Ghai and Wadhi 1983), and the Philippines (Rimando 1962), etc.

In the Philippines, ***Pritchardina fijiensis*** Hirst and ***O. velascoi*** Rimando (Gabriel 1975 *in* Cayme and Gapasin 1987) have also been reported on coconut. The latter species has been particularly closely studied. Its development cycle lasts less than a week and the number of other hosts is small (Cayme and Gapasin 1987).

In the genus ***Tetranychus***, we would also mention ***T. piercei*** Mc Gregor on *Elaeis* in Indonesia (Mariau *et al.* 1991), ***T. mexicanus*** Mc Gregor in South America on *Cocos* (Moura and Vilela 1986), but also on *Elaeis* (Genty *et al.* 1978), ***T. neocaledonychus*** Andra on coconut in Mauritius (Moutia 1958), ***T. piercei*** on *Elaeis* (Rajaratnam and Hock 1975) and ***T. tumidus*** Banks on *Cocos* in Florida (Liu and Tsaï 1998).

The following species have also been reported in the genus ***Oligonychus: O. plegas*** Baker and Pritchard (Mariau *et al.* 1991), on *Elaeis* in Indonesia and on coconut in Mauritius (Moutia 1958), ***O. bagdasariani*** Banks on *Elaeis* in South America (Genty *et al.* 1978). ***O. modestus*** (Banks) in Thailand (Baker 1976), ***O. pratensis*** (Banks) in Mauritius (Moutia 1958) and two ***Oligonychus*** sp. in Sri Lanka (CPCRI 1979) are also mentioned on coconut. In Malaysia, a species of the same genus can cause spectacular damage on *Elaeis* in the nursery (Wood 1968).

A few other species of *Tetranychidae* have been observed: ***Eutetranychus enodes*** Baker and Pritchard on *Elaeis* in West Africa (Mariau *et al.* 1981), ***Amenosius*** sp. on *Cocos* flowers in India (CPCRI 1979) and ***Dolycotetranychus*** sp. on *Cocos* in India (Kanagaratnam *et al.* 1981).

TENUIPALPIDAE

This family includes two species that can cause serious damage.
Raoiella indica Hirst on *Cocos* in India, Mauritius and Sri Lanka. At an average temperature of 24°C, the development cycle lasts 18 to 26 days, but can take longer than 1 month at an average temperature of 18°C (Moutia 1958). More recently, the species ***Rarosiella cocosae*** Rimando has been detected in the Philippines (northern Mindanao); it attacks coconut palms of all ages and losses can reach 87% (Poliquit et al. 1997). Lastly, in this family, we would mention ***Brevipalpus phoenicis*** (Geijskes) which is found worldwide and has numerous hosts, including coconut in India, where it has been found on fronds and nuts (Nagesh Chandra and Channa Basavanna 1976).

Factors involved in reducing *Tetranychidae* and *Tenuipalpidae* populations have been studied
It has been seen that temperature can play an active role in fecundity (Sathiamma 1998, Liu and Tsai 1998). Mineral nutrition can also have a notable impact in regulating populations. We would mention the populations of ***T. piercei,*** for which a higher concentration of mites is found on plants which have received little or no boron (Rajaratman and Law Ing Hoch 1975). It has been shown that, apart from phosphorus and calcium, numerous chemical elements play a role in the size of ***R. indica*** populations (Sarkar and Somchoudhury 1989).
Tetranychidae and *Tenuipalpidae* have numerous natural enemies. Populations of these mites are very frequently associated with predatory mites mainly belonging to the family of the *Phytoseiidae,* but whose genera are often unknown. We would mention *Euseius beninensis* Moraes and McMurthy on *Elaeis* in Benin (Moraes *et al.* 1989), *Lasioseius phytoseiodes* Chant. in Colombia on *Cocos* (Cardona 1971), *Amblyseius paroaerialis* Muma and *A. eucalypticus* Gupta on *Cocos* in India (Sathiamma 1996).

Sometimes, phytophagous mite populations are associated with *Steneotarsonemus furcatus* De Léon on *Cocos*, in Colombia (Cardona 1971). The mite, *Typhlodromus caudatus* Berl. is an active predator on ***R. indica*** (Moutia 1958).
Predator mites belong to other families: *Cunaxidae* with *Cunaxa setirostris* (Hermann), *Agistemus* sp. (*Stigmatidae*), *Lasioseius* sp. (*Ascidae*) and *Eupodes* sp. (*Eupodidae*) associated with ***Oligonychus iseilemae*** colonies in India (Sathiamma 1996).

Also among the predators, Coleoptera *Coccinellidae* of the genus *Stethorus* can play an important role. For instance, *S. expectatus* Chazeau and *S. exsultabilis* Chazeau attack *Tetranychidae* in New Guinea (Chazeau 1983). Several species of the same genus are associated with ***Raoiella indica*** in India: *S. parcepunctatus* Weise, *S. tetranychii, S. keralicus* Kapur (Daniel 1976), whilst *S. siphonulus* is common in Malaysia on *Elaeis* (Wood 1968). Lastly, Moutia (1958) mentions *S. vinsoni* in Mauritius and, to a lesser degree, *Oligota pallidicornis* and *Exochomus laeviusculus* Weise.

When there are very severe mite outbreaks, mostly in the nursery, it may be necessary to carry out insecticide applications. With *Tetranychidae,* sulphur gives very good results (Genty *et al.* 1978, Mariau *et al.* 1981, 1991). ***R. indica*** has been involved in various pesticide trials (Sarkar and Somchoudury 1988).

ERIOPHYIDAE (on nuts)

Numerous species of *Eriophyidae* have been observed, primarily on coconut fruits and fronds.

Eriophyes guerreronis Keifer

This is the most important species. Given the seriousness of the damage it causes and its geographical distribution, which is ever increasing, much has been written about it in the literature.

Brief description
As described by Keifer (1965), ***E. guerreronis*** has a very elongated body. That of the female measures 205 to 255 μ in length and 36 to 52 μ wide. The fore section has two pairs of legs and the finely annulate body bears a few long hairs.

Biology
E. guerreronis can be found at all its development stages under the floral parts of nuts. The mite penetrates the fruit very early after pollination, when the first mites appear on the inner surface of the bracts of a few fruits. In a few weeks, virtually all the fruits show at least the first signs of attack (Mariau 1977, Moore and Alexander

1987). However, attacks can occur later. The development cycle lasts around ten days (Mariau 1977, Suarez 1991a). When the floral parts of a severely attacked young nut are raised, the mite populations appear like a white powder containing thousands of individuals.

Damage

At the earliest, the first symptoms occur 3 to 4 weeks after pollination, in the form of triangular white patches, whose base is located near the petals. These patches turn brown and meet up, and in extreme cases they form a sleeve right round the nut.

It has often been written that attacks lead to substantial nut fall. It has been shown, at least under conditions prevalent in Côte d'Ivoire and on West African Tall material, that attacked and healthy coconut palms produce the same number of nuts, all other things being equal (Mariau 1977). Losses are primarily linked to a lower endosperm content in the nuts; the earlier the attack the greater the reduction. On average, it reaches 50% in the most seriously affected nuts (Mariau 1977, Julia and Mariau 1979). In Africa, the most serious losses are recorded in rainy periods, though damage usually occurs in the dry season; indeed, it is in the driest zones that damage is the greatest (Julia and Mariau 1979, Mariau 1986). According to Moore *et al.* (1991), damage increases in line with N levels, but diminishes in line with K levels.

Attacks vary considerably depending on the variety or hybrid. Out of 20 crosses studied at the same site, the West African Tall (WAT) proved to be the most susceptible with an average copra reduction of 16%, whereas the most tolerant hybrid (Spanish Green Dwarf x Polynesia Tall) only had a 5% copra reduction over the same period (Mariau 1986). Losses can vary substantially within the same variety or cross. For instance, with WAT, losses can vary from 9.5 to 18.6% (Mariau 1986). Similar observations were carried out in Cuba on the "Criollo" variety, which is substantially more attacked than the "Indio Verde" variety (Suarez 1991b) and in Trinidad, where round nuts are less attacked than other nuts (Moore and Alexander 1990). Studies in India (Muthiah and Bhaskaran 1999) showed the Seychelles variety to be particularly susceptible.
No precise information is given for losses on a nationwide scale. In Venezuela, Doreste (1968) estimated losses to reach 70%, which seems very excessive. In Mexico they appeared to be less than 10% (Ortega *et al.* 1965) whereas Moore *et al.* (1989) estimated losses of 32% on Saint Lucia. In Benin, coconut palms located on the coast, hence almost continually ventilated, showed very few signs of attack (Mariau and Julia 1970).

Whilst, generally speaking, damage occurs on nuts, ***E. guerreronis*** has been found to attack the spears of coconut palms in the nursery in Brazil. Such attacks can kill the seedling (Arruda 1974).

Geographical distribution

E. guerreronis was reported for the first time in Mexico in 1960 (Ortega *et al.* 1962 *in:* Ortega *et al.* 1965, Hall 1981). However, very similar damage was reported in Colombia as early as 1948 (Bain *in*: Cardona and Sanchez 1971). The existence of damage was confirmed by Cartujano (1963) and that of the mite itself in 1965. ***E. guerreronis*** was then reported to be present in all American and Caribbean countries: in Venezuela in 1967 (Doreste 1968), in Costa Rica at an uncertain date (Schliesske 1988); later in the Caribbean: in Cuba around 1975 (Ortiz and Avila 1975), Trinidad around 1976 (Griffith 1984), Saint-Lucia at an uncertain date (Moore *et al.* 1989), lastly in Puerto Rico in 1977 and in Florida in 1984 (Howard *et al.* 1990). In Africa, it is on the island of Sao Tomé that it seems to have appeared for the first time (OILB 1968 *in* Mariau 1969). On the continent, it is in Benin that it occurred for the first time (Mariau 1969). It was then rapidly reported throughout West Africa, notably in Côte d'Ivoire (Mariau 1977). It also exists in East Africa. It has only recently been seen in Asia, more particularly on the Indian subcontinent (Sathiamma *et al.* 1998). It was observed in 1999 in the Indian states of Kerala, Tamil Nadu and Karnataka, and on the west coast of Sri Lanka (Haq 1999).

Natural enemies

Other species are found associated with ***E. guerreronis***, some of them predators. These predators belong to three families.

Phytoseiidae, with *Amblyseius largoeusis* (Muma), *Neoseiulus mumai* (Denmark), *N. paspalivorus* (De Léon) in Florida and *N. baraki* Athias-Henriot in Puerto Rico (Howard *et al.* 1990); lastly, a species of the genus *Amblyseius* has been observed in Côte d'Ivoire (Mariau 1977).

Tarsonemidae with *Tarsonemus* sp. in Florida and Puerto Rico, *Rhynchotarsomemus* sp. in Florida, *Steneotarsonemus furcatus* (De Léon) in Puerto Rico (Howard *et al.* 1990) and *Lupotarsonemus* spp. in Côte d'Ivoire and Papua New Guinea (Hall *et al.* 1980).

Bdellidae with *Bdella distincta* (Baker and Balogh) in Puerto Rico and *B. indicata* in Côte d'Ivoire. The efficacy of these mites is dubious to say the least, in that they do not seem to have any effect in regulating ***E. guerreronis*** populations.

The other mites found in association with ***E. guerreronis*** belong to the families of the *Acaridae* and *Tydeidae;* the species belonging to these families feed off debris.

Several species of fungus, including *Hirsutella thompsonii*, have been isolated from ***E. guerreronis*** populations. All the strains of this fungus are pathogenic to mites, and the one found in Côte d'Ivoire seems to be the most virulent (Hall *et al.* 1980). Trials testing this strain were positive when populations were treated directly in the laboratory, but the same strain proved to be totally ineffectual when applied in the field (Mariau, unpublished). More encouraging results seem to have been obtained in Mexico (Becerril and Sanchez 1986) and in Cuba (Cabrera and Dominguez 1987). The strain *H. thompsonii* var. *synnematosa* has been discovered in India (Beevi *et al.* 2000).

Chemical control

Chemical control trials were conducted in Mexico (Ortega *et al.* 1965) and in Africa (Mariau and Tchibozo 1973, Mariau 1977). It was shown that there was no direct effect on mites, which remain well protected under the floral parts. Nevertheless, when treatments were repeated frequently, there was a substantial reduction in damage, particularly with chinomethionate and tricyclohexyl tin hydroxide (Mariau 1977). Due to the persistence of attacks, treatments have to be carried out at such a frequency (at least monthly) that they are not cost-effective in commercial plantations, except in particular cases, such as seed garden protection.

Colomerus novahebridensis Keifer

This common mite in Asia and the Pacific has a very similar behaviour to ***E. guerreronis*** (Kang 1981). However, for unknown reasons, attacks do not cause any serious damage on nuts.

Amrineus cocofolius Flechtman

Recently discovered on coconut in Brazil, causing similar damage to that of ***E. guerreronis*** (EMBRAPA 2000).

ERIOPHYIDAE **(on leaves)**

Retracrus elaeis Keifer was reported for the first time in 1974 on oil palm fronds in Colombia (Genty and Reyes 1977). The female is 154 to 166 µ long and 62 µ wide. The body is covered with an abundant waxy protection. The punctures made by the mite, starting on the lower fronds, lead to orangey discolorations that can affect entire fronds or, in the event of very large populations, virtually the entire leaf area of the palm over thousands of hectares, resulting in production losses of 50%. Some *Elaeis guineensis* x *E. oleifera* crosses are highly susceptible to atttacks, others are much more tolerant or even totally resistant.

Biological control of this mite has been attempted. Application of solutions containing the fungus *Hirsutella thompsonii* Fisher have proved effective (Urueta 1980). However, sulphur spraying is preferred, as it gives very good results and is ecologically acceptable (Genty and Reyes 1977, Mariau and Genty 1987).

Three species of *Eriophyidae* have been observed on oil palm fronds in Côte d'Ivoire: ***Tegonotus gutierrezi*** Boczek and Natcheff, ***Phillocoptes mariaui*** Boczek and Natcheff and ***Epitrimerus elaeis*** Boczek and Natcheff (Boczek and Natcheff 1989). These mites only cause unconsequential damage.

Four other species have been reported on coconut fronds in the Philippines: ***Acathrix trymatus*** Keifer, ***Scolocenus spiniferous*** Keifer, ***Dialox stellatus*** Keifer and ***Notostrix attenuata*** Keifer. These species have been suspected of being vectors of *cadang-cadang* disease, though no proof has been found (Briones 1963). Other species have been reported on coconut ***Acamina coconuciferae*** Keifer, ***Nacerimina jamaica*** Keifer in South America and ***N. gutierrezi*** Keifer in Asia (Keifer *in* Schliesske 1988).

ORIBATIDAE

We would mention, for the record, four species observed on coconut in India. They are species exclusively feeding off debris: ***Uracrobates indicus, Siculobata malabarica, Notogalumna nortoni*** and ***Afronothrus arboreus*** (Ramani and Haq 1990, 1992, 1998).

PYEMOTIDAE

A single species belonging to this family is associated with coconut. *Pyemotes ventricosus* Newport. It is a very active predator on *Promecotheca coeruleipennis* Blanchard (Col. *Chrysomelidae*) and *Levuana iridescens* Bethune-Baker (Lep. *Zygaenidae*) in Fiji where, due to its extremely high multiplication rate, it can annihilate populations of some pests, but also of their parasitoids, which consequently results in spectacular outbreaks of *P. coeruleipennis* in particular. It has also been reported in India as a predator on *Nephantis serinopa* Meyrick (*Opisina arenosella* Walker) (Lep. *Oecophoridae*) (Lever 1969).

MITES REFERENCES

Arruda G.P. de (1974). Fenologia de *Eriophyes guerreronis* (Keifer 1965) (Acarina, Eriophyidae) em Pernambuco. Estado de Permambuco Secretaria de Agricultura Instituto de Pesquisas Agronomicas. *Boletim Técnico* n 66, 54 p.

Baker E.W. (1976). Some spider mites from Southeast Asia and Japan. *FAO, SEA and P.R. Techn. Doc.* N 103, 11 p.

Becerril A.E., Sanchez J.L.C. (1986). El hongo *Hirsutella thompsonii* Fisher en el control del eriofido del cocotero, *Eriophyes guerreronis* (Keifer). *Agric. Téc. Mex.* 12 (2): 319-323.

Beevi S.P., Beena S., Lyia K.R., Varma A.S., Mathew M.P., Nadarajan L. (2000). Report of pathogen, *Hirsutella thompsonii* va. *synnematosa* Samson, McCoy & O'Donnell on coconut mite *Aceria (Eriophyes) guerreronis* (Keifer). *Insect Environment* 5 (4): 190-191.

Boczek J., Natcheff P. (1989). Three new species of Eriophyid mites (Acari: Eriophyoidea) from oil palms in Ivory Coast. *Acarologia* 30 (3): 249-253.

Briones M.L. (1963). Habitat, morphologie sommaire et répartition géographique de quatre nouvelles espèces de phytoptes découvertes sur le cocotier aux Philippines. *Bull. Phyto. FAO.* 11 (2): 25-30.

Cabrera R.I., Dominguez D. (1987). El hongo *Hirsutella nodulosa*, nuevo parasito para el acaro del cocotero *Eriophyes guerreronis*. *Cienc. Tec. Agric. Citricos y Otros Frutales* 10 (1): 41-51.

Cardona I.Z. (1971). Lista preliminar de acaros de importancia en Colombia. *Acta Agronomica Colombia* 21 (3): 119-132.

Cardona Z.I., Sanchez A.P. (1971). La roña o escoriación de lo frutos de cocotero (*Cocos nucifera* L.) en Colombia. *Oléagineux* 26 (12): 767-770.

Cartujano F. (1963). El cocotero en Mexico. En Seminarias de otono, Escuela nacional de agricultura, Chapingo (Mexico): 57-92.

Cayme T., Gapasin D. P. (1987). Biology, hostrange and natural enemies of the coconut spider mite, *Oligonychus velascoi* Rimando. *Annals of Tropical Research* 9 (2): 59-67.

Channa Bassavana G.P., Banu K. (1972). Plant feeding mites of India 3 spider mites of the species *Oligonychus* (Acarina: Tetranychidae). *Mys. J. Agric. Sci.* 6: 163-168.

Chazeau J. (1983). Deux prédateurs de *Tetranychidae* en Nouvelle-Guinée: *Stethorus expectatus* n. sp. et *Stethorus exsultabilis* n. sp. (Col.: Coccinellidae). *Entomophaga* 28 (4): 373-378.

CPCRI (1979). Nematodes, fungi, insects and mites associated with the coconut palm. Central Plantation Crops Research Institute Kasaragod, Kerala, India, *Techn. Bull.* 2, 236 p.

Daniel M. (1976). Biology and predatory habits of the lady bird beetle *Sthethorus keralicus* Kapur (Coleoptera: Coccinellidae), predatory on the palm mite. *Journal of Plantation Crops* 4 (1): 7-9.

Doreste E.S. (1968). El acaro de la flor del cocotero (*Aceria guerreronis* Keifer) en Venezuela. *Agronomia Tropical* 18 (3): 379-386.

EMBRAPA (2000). Mancha anelar do fruto do coquiero. Centro de Pesquisa Agropecuária dos Tabuleiros costeiros Aracaju Se.

Genty Ph., Desmier de Chenon R., Morin J.P. (1978). Ravageurs du palmier à huile en Amérique latine. *Oléagineux* 33 (7): 326-419.

Genty Ph., Reyes E. (1977). Un nouvel acarien du palmier à huile: l'Eriophyidae *Retracrus elaeis* Keifer. *Oléagineux* 32 (6): 255-262.

Ghai S., Wadhi S.R. (1983). Phytophagous mites on coconut. *Exotic Plant Quarantine Pests and Procedures for introduction of Plant materials*: 87-96.

Griffith R. (1984). The problem of the coconut mite, *Eriophyes guerreronis*, in the coconut groves of Trinidad and Tobago. *Proc. Caribb. Fond Crops Soc.* 20: 128-132.

Gutierrez J. (1974). Species of the genus *Tetranychus* Dufour (Acarina: Tetranychidae) of economic importance in Madagascar and the neighbouring island. Competition between the complexe *Tetranychus neocaledonicus* Andre and *Tetranychus urticae* Koch. *Acarologia* 16 (2): 258-270.

Hall R.A. (1981). The coconut mite, *Eriophyes guerreronis*, with special reference to the problem in Mexico. *Proc. Br. 1981 Crop Prot. Soc*: 113-119.

Hall R.R., Hussey N.W., Mariau D. (1980). Results of a survey of biological control agents of the coconut mite *Eriophyes guerreronis*. *Oléagineux* 35 (8-9): 395-400.

Haq M.A. (1999). Distribution of the coconut mite *Aceria guerreronis* in peninsular India and adjacent islands. *Entomon.* 24 (4): 371-379.

Howard F.W., Abreu. Rodriguez E., Denmark H.A. (1990). Geographical and seasonal distribution of the coconut mite, *Aceria guerreronis* (Acari: Eriophyidae), in Puerto Rico and Florida, USA. *J. Agric. Univ. P.R.* 74 (3): 237-251.

Julia J.F., Mariau D. (1979). Nouvelles recherches en Côte d'Ivoire sur *Eriophyes guerreronis* K, acarien ravageur des noix du cocotier. *Oléagineux* 34 (4): 181-189.

Kanagaratman P., Pinto J.L., Sinnathamby S.V. (1981). Some minor pest of coconut; new records for Sri Lanka. *Ceylan Cocon. Quat.* 32 (3-4): 93-95.

Kang S.M. (1981). Eriophyides et tarsonémides chez le cocotier. *FAO Bulletin Phytosanitaire* 29 (3-4): 79-80.

Keifer H.H. (1965). Eriophyd studies B. 14. *Bureau of Entomology California, Department of Agriculture*: 7-8.

Lever R.J.A.W.(1969). Les ravageurs du cocotiers. *Etudes agricoles de la FAO.* n 77, 196 p.

Liu Y.H., Tsai J.H. (1998). Development, survivorship, and reproduction of *Tetranychus tumidus* Banks (Acarina Tetranychidae) in relation of temperature. *Internat. J. Acarol.* 24 (3): 245-252.

Mariau D. (1969). *Aceria guerreronis* Keifer: récent ravageur de la cocoteraie dahoméenne. *Oléagineux* 24 (5): 269-272.

Mariau D. (1977). *Aceria (Eriophyes) guerreronis*: un important ravageur des cocoteraies africaines et américaines. *Oléagineux* 32 (3): 101-111.

Mariau D. (1986). Comportement de *Eriophyes guerreronis* Keifer à l'égard de différentes variétés de cocotiers. *Oléagineux* 41 (11): 499-505.

Mariau D. (1989). Les acariens inféodés au palmier à huile et au cocotier. Colloque sur les acariens des cocotiers. Montpellier, octobre 1989. *Annales A.N.P.P.* 2 (1): 239-247.

Mariau D., Desmier de Chenon R., Julia J.F. (1981). Les ravageurs du palmier à huile et du cocotier en Afrique de l'Ouest. *Oléagineux* 36 (4): 168-228.

Mariau D., Desmier de Chenon R., Sudharto P.S. (1991). Les ravageurs du palmier à huile et leurs ennemis en Asie du Sud-Est. *Oléagineux* 46 (11): 400-476.

Mariau D., Genty P. (1987). Utilisation du soufre dans la lutte contre *Retractus elaeis* Keifer, acarien foliaire du palmier à huile en Colombie. Int. symp. elemental sulphur in agriculture. Agropolis Nice, France.

Mariau D., Julia J.F. (1970). L'acariose à *Aceria guerreronis* (Keifer) ravageur du cocotier. *Oléagineux* 25 (8-9): 459-464.

Mariau D., Tchibozo H.M. (1973). Essais de lutte chimique centre *Aceria guerreronis* (Keifer). *Oléagineux* 28 (3): 133-135.

Moore D., Alexander L. (1987). Aspects of migration and colonization of the coconut palm by the coconut mite, *Eriophyes guerreronis* (Keifer) (Acari: Eriophyidae. *Bull. Ent. Res.* 77: 641-650.

Moore D., Alexander L. (1990). Resistance of coconuts in St Lucia to attack by the coconut mite *Eriophyes guerreronis* Keifer. *Trop. Agric. (Trinidad)* 67 (1): 33-36.

Moore D., Alexander L, Hall R.A. (1989). The coconut mite, *Eriophyes guerreronis* Keifer in St Lucia: yield losses and attempts to control in with acaricide, polybutene and *Hirsutella* fungus. *Tropical Pest management* 35 (1): 83-89.

Moore D., Ridout M.S., Alexander L. (1991). Nutrition of coconuts in St Lucia and relationship with attack by coconut mite *Aceria guerreronis* Keifer. *Trop. Agric. (Trinidad)* 68 (1): 41-44.

Moraes G.J. de, Mc Marthy J.A., Yaninek J.S. (1989). Some phytoseiid mites (Acari: Phytoseiidae) from tropical Africa with description of a new species. *Internat. J. Acarol.* 15 (2): 95-102.

Moura J.I.L., Vilela E.F. (1996). Insect pests of coconut palm and oil palm. *Imp. Jard. Viçosa Minaisgerais*, Brazil 73 p.

Moutia L.A. (1958). Contribution to the study of some phytophagous acarina and their predators in Mauritius *Bull. Ent. Res.* 49: 59-75.

Mutiah C., Bhaskaran R. (1999). Screening of coconut genotypes and management of eriophyid mite *Aceria guerreronis (Eriopyidae: Acari)* in Tamil Nadu. *Indian Coconut J.* 30 (1): 10-11.

Nagesh Chandra B.K., Channa Basavanna G.P. (1976). Host plants of *Brevi palpus phoenicis* (Geijskes) (Acarina: Tenuipalpidae) in India. *Acarology Newsletter* n 2: 3.

Ortega A., Rodriguez J., Caribay C. (1965). Investigaciones preliminares sobre el eriofido del cocotero, *Aceria guerreronis* Keifer, en la Costa Grande de Guerrero. *Agricultura Tecn. Mexico* 2 (5): 222-226.

Ortiz J.E., Avila M.G. (1975). Daños causados en coco por *Aceria guerreronis* (Acarina: Eriophyidae) en Cuba. *Rev. de Agricultura* 8 (2): 30-34.

Poliquit E.S., Escalona R.T., Sabornido E. (1997). *Rarosiella cocosae* Rimando, a new mite infesting coconut palms in northern Mindanao, Philippines. *PJ.C.S.* 12 (1): 50-53.

Rajaratnam J.A., Law Ing Hock (1975). Effect of boron nutrition on intensity of red spider mite attacks on oil palm seedlings. *Experimental Agriculture* 11 (1): 59-63.

Ramani N., Haq M.A. (1990, 1990, 1992, 1998). Oribatid mites from coconut palm: 1. A new species of *Uracrobates* (Acari: Oribatei) from Kerala (India). *Acarologia* 31 (3): 297-302; 2. A new species of *Notogalumna* (Acari: Oribatei) from Kerala (India). *Acarologia* 31 (4): 401-404; 3. A new species of *Afronothrus* (Acari: Oribatei: Trypochthoniidae) from Kerala (India). *Acarologia* 33 (2): 207-212; 4. A new species of *Siculobata* (Acari: Oribatei) from Kerala, India. *Acarologia* 39 (1): 85-90.

Rimando L.C. (1962). The tetranychoid mites of the Philippines. *Univ. Phil. Coll. Agric. Tech. Bull.* 11: 11-12.

Sarkar P.K., Somchoudhury A.K. (1988). Evaluation of some pesticides against *Raoiella indica* Hirst on coconut palm in West Bengal. *Pesticides* oct. 1988: 21-22.

Sarkar P.K., Somchoudhury A.K. (1989). Interrelationship between plant characters and incidence of *Raoiella indica* Hirst. on coconut. *Ind. J. Ent.* 51 (1): 45-50.

Sathiamma B. (1988). Record on the red spider mite *Tetranychus ludeni* Zacher (Acarina: Tetranychidae) on the coconut palm. *Entomon.* 13 (2): 191-192.

Sathiamma B. (1996). Observations on the mite fauna associated with the coconut palm in Kerala, India. *Journal of Plantation Crops* 24 (2): 92-96.

Sathiamma B. (1998). Screening of plant species for susceptibility to infestation by *Oligonychus iseilemae* (H.) (Tetranychidae, Acarina) the white spider mite of coconut palm. *Journal of Plantation Crops* 26 (1): 80-82.

Sathiamma B., Nair C.P.R., Koshy P.K. (1998). Outbreak of a nut infesting eriophyid mite *Eriophyes guerreronis* (K.) in coconut plantations in India. *Indian Coconut J.* 30 (2): 1-3.

Schliesske J. (1988). Zur Gullmilbenfauna (Acari: Eriophyoidea) von *Cocos nucifera* L. in Costa Rica. *Nachrichtenbl. Deut. Pflanzenschutzd.* 40 (8-9): 124-127.

Suarez A. (1991a). Estudios biologicos de *Eriophyes guerreronis* (Acarina, Eriophyidae) en condiciones de Laboratorio. *Proteccion de Plantas* 1 (2): 47-51.

Suarez A. (1991b). Distribucion y evaluacion de daños del acaro *Eriophyes guerreronis* en plantaciones de cocotero de la region de Baracoa. *Proteccion de Plantas* 1 (1): 75-81.

Urueta E.J. (1980). Control del acaro *Retracus elaeis* Keifer mediante el hongo *Hirsutella thompsonii* Fisher e inhibicion de este por dos fungicidas. *Augura* Medellin Colombia 6 (2): 25-31.

Wood B.J. (1968). Pests of oil palm in Malaysia and their control. The incorporated society of planters, Kuala Lumpur, Malaysia, 204 p.

GENERAL CONCLUSION

It is in the order of the Lepidoptera, at the larval stage, that the largest numbers of oil palm and coconut pests are to be found, and a particularly large number of species in the family of the *Limacodidae.* Lepidoptera are mostly leaf-eaters, though a few species mine galleries in different parts of the plant.

There are also major pests in the order of the Coleoptera, more particularly in the families of the *Scarabaeidae Dynastinae,* which cause damage as adults, the *Curculionidae*, whose larvae mine galleries in the plant, and the *Chrysomelidae Hispinae,* leaf-eaters in both the adult and larval stages. The species belonging to the last family are leaf browsers or miners.

Numerous very serious pests can be found among the Hemiptera, sucking insects including numerous primary pests and vectors of viral, phytoplasma and protozoan diseases, among the Orthoptera, including a few major leaf-eaters, notably in Asia, among the Isoptera, which often attack living palms, and finally among the mites, which puncture leaves and fruits.

Without natural factors to regulate their populations, there would be virtually continual outbreaks of many of these pests. The biotic factors that are antagonistic to crop pests mainly occur among insect and mite parasitoids or predators, along with diseases, primarily caused by fungi and viruses.

All this fauna and flora in a palm plantation combines to form a highly complex whole, in which very cautious intervention is required when pest populations increase abnormally. Otherwise, durable imbalances might be created, resulting in repetitive outbreaks that are sometimes difficult to curb.

If a pest outbreak is to be effectively controlled in a balanced manner, it is first of all necessary to be familiar with the fauna in the palm planting. That is one of the aims of this work which, with approximately 800 references, reviews all the actual and potential pests, along with biotic agents that usually guarantee population stability.

INDEX

Abgrallaspis cyanophylli,90, 91
Acamina coconuciferae,204
Acaria,20
Acaridae,202
Acarina,150
Acathrix trymatus,204
Acraga ochracea,25
Acrididae,178
Acridoidea,4, 177, 178, 181, 182, 183
Acritocera negligens,21
Actia painei,13
Acysta interrupta,93
Adelencyrtus oceanicus,91
Adoretus,110, 111
Adoretus borneensis,110
Adoretus celogaster,111
Adoretus compressus,110
Adoretus dorsalis,111
Adoretus lasiopygus,111
Adoretus lithobius,111
Adoretus sinicus,111
Adoretus umbrosus,111
Afronothrus,209
Afronothrus arboreus,204
Agathis,12
Agenius colini,111
Agenius quadrimaculatus,111
Agistemus,200
Agonoxena,10, 64, 66, 70, 78, 80, 82
Agonoxena argaula,12, 13, 64, 70, 78, 80, 82, 187
Agonoxena miniana,13
Agonoxena pyrogramma,13, 66
Agonoxeninae,12
Ahasverus,171
Ahasverus advena,172
Alaucapsis,91
Alaus lacteus,172
Alcorrhyncus grandis,27, 57, 95
Aleiodes,25, 31
Aleurocanthus,88
Aleurocanthus cocois,88
Aleurocanthus dispersus,88

Aleurocanthus flavus,88
Aleurocanthus gateri,88
Aleurocanthus neglectus,88
Aleurocanthus pulvinatus,88
Aleurocantus,89
Aleurodicus destructor,88, 107
Aleuronudus,88
Aleyrodidae,88, 98, 103, 107
Aleyrodoidea,3, 88
Allecula,8
Allecula minuta,8
Alleculidae,8
Allodontermes morogorensis,188
Alurnini,130, 132
Alurnus humeralis,132, 136, 145
Amathusia phidippus,55
Amathusiidae,55
Amatissa cuprea,9
Ambadra rafflesi,61
Amblypelta cocophaga,92, 196
Amblyseius,199, 202
Amblyseius eucalypticus,199
Amblyseius largoeusis,202
Amblyseius paroaerialis,199
Amenosius,199
Amerris (Amerrhinus) ynca,149
Amerris parcus,149
Aminellus indicus,90
Amitermes evuncifer,188
Amitermes globosus,188
Amitermes sulphureus,188
Amorphoidea,147, 167
Amorphoidea coimbatorensis,147
Amphiacusta caraiba,180
Amrineus cocofolius,203
Amsacta transiens,61
Amyotea,136
Amyotea antigua,136
Amyotea reciproqua,136
Anastatus,53, 138
Ancylorrhynchus trapezicollis,147
Aneuradesha harleyi,135
Anigrus,86
Ankylopteryx octopunctata candida,17, 79
Anobiidae,173
Anocofana mukilengensis,87
Anoplolepis longipes,185, 196, 197
Antaeotricha,14

Antaeotricha enthruca,14
Antaeotricha phaenora,14
Anthocoridae,17
Anthribidae,173
Antigestra,51
Antrocephalus,31, 51, 63, 74
Antrocephalus cariniceps,19
Antrocephalus hakonensis,16, 19, 63, 74
Antrocephalus phaeospilus,19
Antrocephalus renalis,48
Aonidiella cremocitri,90
Aonidiella orientalis,90, 91
Apanteles,8, 9, 10, 12, 14, 18, 23, 27, 28, 31, 32, 34, 35, 37, 40, 50, 52, 57, 60, 82
Apanteles agilis,53
Apanteles agonoxenae,12, 13
Apanteles alius,57
Apanteles aluella,27, 30, 31, 32, 33, 40
Apanteles anyaleti,8
Apanteles artonae,23
Apanteles belippae,31
Apanteles caniae,30
Apanteles hidaridis,53
Apanteles metesae,8, 9
Apanteles orelia,12
Apanteles parasae,23, 27, 35, 36, 38
Apanteles penthocratus,27, 36, 37
Apanteles pertiades,49
Apanteles psychidivorus,9
Apanteles roepki,40
Apanteles sp. nr. glomeratus,39
Apanteles stantoni,13
Apanteles taragmae,16, 77
Apanteles targamae,19
Apanteles tirathabae,48, 49
Apha manilae,19
Aphamisticus altus,170
Aphanocorynes humeralis,155
Aphanogmus,13, 40, 67
Aphanogmus goniozi,16, 19, 67
Aphanogmus hakonensis,8
Aphanogmus manilae,19, 29, 32, 33, 35, 36, 39, 40, 48, 49
Aphareta,36
Aphelinidae,90, 138
Aphididae,89, 105
Aphidoidea,3, 89
Aphytis chrysomphali,90, 106
Apidae,196

Apleurotropis lalori,138
Apogonia expeditionis,111
Aposporium,88
Araecerus fasciculatus,173
Arciera grisea,61
Arctiidae,61
Argyrophylax basifulva,49, 69
Aroplectrus,28
Aroplectrus dimerus,32, 36, 37
Artona catoxantha,17, 22, 23, 72, 75, 82
Artona. catoxantha,23
Ascidae,200
Ascyltus pterygotes,184
Asota plana,61
Aspergillus,8
Aspergillus candidus,93
Aspergillus flavus,47
Aspidiotini,90
Aspidiotus,90, 95, 97, 98, 101, 102, 106
Aspidiotus destructor,89, 90, 95, 97, 98, 101, 102, 106
Aspidomorpha cincta,129
Aspidomorpha obovata,129
Astegopteryx,89, 107
Astegopteryx nipae,89, 107
Astegopteryx rappardi,89
Astegopteryx rhaphidis,89
Asterolecaniidae,92, 103
Asura biseriata,61
Atractomorpha psittacina,179
Atta cephalotes,196, 197
Augosoma centaurus,118, 126
Aularches miliaris,179
Aulosaphes,8
Aulosaphes psychidivorus,8, 9
Aulostyrax nuciferae,131
Austrophorocera,40
Austrophorocera grandis,29, 31, 34, 40
Automeris,52
Automeris bilinea,52
Automeris cinctistiga,52
Automeris janus,52
Automeris liberia,52
Automeris vividior,52
Axiagastus cambelli,93
Azteca,196, 197
Baculovirus,114, 115
Baculovirus oryctes,117, 118
Bakera nigroscuta,87

Balanephagus attaleae,147
Balyana,135
Balyana mariaui,135
Baria elsa,30, 45
Barichneumon muciallae,48
Baridinae,147
Batotheca,31
Batrachedra arenosella,14, 64
Batrachedra nuciferae,14
Batrachedra perobtusa,14
Bdella distincta,202
Bdella indicata,202
Bdellidae,202
Beauveria,43
Beauveria bassiana,47, 56, 117
Beauveria brongniarti,14
Bellicositermes nigeriensis,188
Belvosia formosa,52
Bessa (Ithychomyia) remota,23
Bessa remota,17, 23, 29, 32, 36, 71, 76
Bethylidae,9, 16, 18, 19
Bethyloidea,28
Birthamoides circinatus,30
Birthamula chara,30
Birthosea bisura,30, 45
Bitoma palmarum,172
Blaesoxipha filipjevi,180
Blastobasidae,15
Blastobasis byrsodepta,15
Blastobasis chloroptris,15
Blepharella lateralis,36
Boettcherisca peregrina,31
Bombycoidea,3, 52
Bombyliidae,29, 30, 69
Bothryperus psaltes,170
Botryonopa sanguinea,132
Botryonopini,130, 132
Brachycoryphus nursei,16, 20, 77
Brachymeria,9, 10, 12, 13, 16, 18, 23, 28, 34, 38, 40, 41, 53, 54, 57, 77, 136
Brachymeria agonoxenae,12, 13
Brachymeria apicicornis,23
Brachymeria carinata,9
Brachymeria euploeae,23, 53
Brachymeria exacarinata,19
Brachymeria fijiensis,13
Brachymeria ghani,8
Brachymeria hammari,13

Brachymeria hime atteviae,19
Brachymeria lasus,16, 19, 28, 31, 32, 33, 35, 36, 40, 77
Brachymeria lasus,,38
Brachymeria lugubris,8, 23
Brachymeria mnestor,18
Brachymeria nephantidis,16, 17, 19, 20, 77
Brachymeria nosatoi,16, 19
Brachymeria nr lasus,58
Brachymeria nr. jambolana,58
Brachymeria obscurata,53
Brachymeria polynesialis,13
Brachymeria punctiventris,23
Brachymeria subconica,18
Bracon,10, 12, 13, 19, 71, 79, 83, 136
Bracon brevicornis,16, 19, 71, 83
Bracon hebetor,16, 79
Bracon serinopae,19
Braconidae,8, 9, 10, 12, 13, 14, 16, 17, 18, 19, 20, 23, 25, 27, 28, 29, 30, 48, 52, 54, 57, 60, 61, 63, 79, 82, 134, 135, 145, 154
Brassolidae,56
Brassolis,56, 64, 70
Brassolis astyra astyra,56
Brassolis isthmia,56, 64
Brassolis sophorae,56, 70
Brevicolaspis villosa,129
Brevipalpus phoenicis,199
Brontispa,130, 131, 137, 139, 140, 143, 144
Brontispa castaneipennis,131
Brontispa chalybeipennis,131
Brontispa depressa,131
Brontispa limbata,131
Brontispa longissima,130, 131, 136, 139, 140
Brontispa longissima celebensis,130
Brontispa longissima frogatti,130
Brontispa longissima javana,130
Brontispa longissima longipennis,130
Brontispa mariana,131
Brontispa palauensis,131, 144
Brontispa suriganoa,131
Bruchidae,173, 174, 176
Buprestidae,170
Buysmania oxymora,27, 30, 31, 32, 33, 34, 35, 36, 37, 38
Caccoleptus wicki,57
Cadra cautella,173
Cadurcia leefmansi,23
Calenodispa sarrasini,131
Caligo nr. eurilochus,57
Caligo oedipus fruhstorferi,57

Caligoninae,57
Calleida splendida,17
Callimerus arcufer,8, 23
Callimerus bellus,8
Calliphoridae,53
Callispa,131, 144
Callistola tripartita,131
Calliteara horsfieldii,60
Calotermes,192
Calotermes (Neotermes) rainbowi,190, 192
Calotermes immigrans,190
Calotermes laticollis,190
Calyptocephala brevicornis,129, 144
Campoletis,56
Camponotus,93
Cania bandura,30
Cania robusta,30
Cania siamensis,31
Cania striola,31
Cantheconidea,8, 95
Cantheconidia furcellata,71
Carabidae,13, 17, 23, 68, 76, 113
Cardiastethus exigruis,17
Cardiodactylus novae-guineae,180
Carinichremylus peleopodae,18
Carpophilus,171
Carpophilus dimidiatus,171, 172
Carpophilus foveicollis,171
Carpophilus hemipterus,171, 172
Carpophilus humeralis,171
Carpophilus nitidus,171
Carvalhoia arecae,93, 99
Casinaria,28, 34, 35, 39
Casphalia extranea,31, 45, 67
Cassida,129
Cassidinae,128, 144
Castnia daedalus,21, 22, 70, 72, 73
Castnia licus,22
Castniidae,21, 70, 73, 78
Catantops humilis,179
Catantops splendens,179
Catollerus strangularis,147
Cecidomyiidae,148
Cedusa,85
Celestes,147
Centrodora,138
Cephaloleia,132, 145
Cephaloleia depressa,132

Cephaloleia elaeidis,132
Cephaloleia nr. vagelineata,132
Cephaloleini,130, 132
Cephrenes chrysozona,54
Ceraleurodicus,88
Cerambycidae,110, 170, 174
Ceraphron,20, 29
Ceraphron fijiensis,49
Ceraphron) fijiensis,13
Ceraphronidae,8, 19, 20, 29, 39, 48, 64
Ceraphronoidea,28
Cerataphis,101, 105
Cerataphis brasilensis,89
Cerataphis lataniae,89
Cerataphis variabilis,89, 101
Ceratopoginidae,29
Cercopidae,87
Ceresa cuprea,87
Ceresa vitulus,87
Ceromasia sphenophor,154
Ceromya rotundicornis,40
Ceroplastes,91
Cetoniinae,110, 111
Chaetexorista javana,29, 30, 31, 32, 33, 35, 36, 38, 39, 40, 41
Chaetexorista javana,32
Chalcididae,8, 9, 12, 13, 16, 18, 19, 23, 24, 25, 28, 30, 48, 49, 50, 51, 52, 53, 54, 55, 56, 57, 58, 63, 74, 75, 136, 140, 170
Chalcidoidea,28, 38, 65, 72, 142
Chalcocelis albiguttatus,29, 31, 47
Chalconicles catori,64
Chalcoscelides castaneipars,31
Chalcosoma atlas,119
Chanteconidea,27
Chanteconidea furcellata,27
Cheilomenes sulfurea orbicularis,90
Cheilomenes vicina,90
Cheiranthium,18
Chelonus,14, 64
Cheromettia,26
Cheromettia sumatrensis,31
Chevalierella elaeidis,51
Chilochorus cacti,90
Chilochorus circumdatus,90
Chilochorus dohrni,90
Chilochorus nigritus,90
Chilochorus schiodtei,90
Chlorocryptus purpuratus,27, 30, 38, 39, 40, 41
Cholus parcus,149

Cholus zonatus,149, 164
Chrysididae,28, 31, 40, 77
Chrysis,31, 32, 34
Chrysis shanghaiensis,40
Chrysis spina,34
Chrysomelidae,3, 110, 128, 129, 136, 139, 140, 141, 142, 143, 144, 145, 204, 211
Chrysomphalus aonidum,91
Chrysomphalus ficus,90
Chrysonotomyia,135
Chrysophilus ferruginosus,154
Chrysopidae,17
Cicadellidae,84, 87, 98, 109
Cicadelloidea,87
Cicadidae,87
Cicadomorpha,84, 95
Cirrospilus,138
Cirtomenus bergi,94, 104
Cixiidae,85, 97, 99, 100, 101, 103, 108
Clania,9
Clania bifenestralis,9
Clania tertia,9
Cleridae,8, 172, 174
Clinocentrus,36
Closterocerus,132, 135
Closterocerus africanus,134
Closterocerus leptocerus,134
Closterocerus promecothecae,138
Closterocerus splendens,138
Clostoptera undulata,87
Clovia,87
Clubionidae,18, 79
Coccidae,91, 106, 107
Coccinellidae,100, 106, 107, 200, 206
Coccoidea,3, 89
Coccus hesperidum,92
Coconympha iriarcha,15
Cocotrypes,169
Cocotrypes congonus,170
Cocotrypes dactyliperda,170
Cocotrypes nigripes,170
Cocotrypes perditor,170
Cocygodes coccys,34
Coelaenoenodera perrieri,135
Coelaenomenodera,128, 133, 135, 139, 140, 141, 142, 143, 144, 145
Coelaenomenodera elaeidis,133, 134, 139, 140, 143, 144, 145
Coelaenomenodera lameensis,128, 133, 134, 135, 136, 137, 139, 143
Coelaenomenodera minuta,133, 141, 144, 145

Coelaenomenodera perrieri,135
Coelaenomenodera speciosa,133
Coelaenomenoderini,130
Colaspis,129
Coleoneura trichogramma,49
Coleoptera,3, 57, 110, 189
Colomerus novahebridensis,203
Colydiidae,172
Compsilura concinnata,29
Conidiobolus,190
Contheyla rotunda,31, 47, 74, 75
Contigucephalus,85
Conura,18, 28, 34, 56, 132, 136
Conura (Conura) papo,28
Conura adela,132
Conura biannulata,34, 39
Conura camescens,39
Conura celsa,18
Conura desmieri,136
Conura elaeisis,10
Conura hispinephaga,136
Conura immaculata,18, 28, 34, 35, 39
Conura martinezi,28, 37
Conura morleyi,56
Conura phobetronae,28, 37
Coptotermes,191, 192
Coptotermes ceylonicus,189
Coptotermes curvignathus,189, 190, 192
Coptotermes heimi,189, 190
Coptotermes niger,189
Coptotermes sinabangensis,190
Coptotermes travians,189
Coptotermes truncatus,189
Coraliomela,132, 141
Coraliomela brunnea,132, 141
Coraliomela marginata,132
Coraliomela quadrimaculata,132
Corcelia rufiventris,30
Corcyra cephalonica,51, 173
Cordyceps,43, 44, 47, 114
Coreidae,92
Corvus validus,178
Cosmolestes picticeps,8, 27
Cosmopterygidae,14
Cossidae,21, 70
Cossoidea,3, 21
Cossoninae,155
Cotesia,27, 28, 39, 52, 57

Crematogaster,93
Crematogaster gabonensis,134
Crematopsyche pendula,9, 72
Cryptinae,27, 52, 136
Cryptognatha nodiceps,90
Cryptonichini,130, 131
Cryptonychus cryptonychellus breviceps,131
Cryptonychus cryptonychellus breviceps palmarum,131
Cryptorrhynchinae,147
Cryptothelea cardiophaga,9
Ctenochiton japonicus,92
Ctenocompa hilda,31
Cucujidae,171, 172
Cucujoidea,171
Cunaxidae,200
Curculionidae,3, 110, 147, 156, 157, 158, 159, 160, 161, 162, 163, 164, 165, 166, 167, 168, 171, 194, 211
Cyclodes omma,59
Cydnidae,94
Cydostethus,154
Cyrtacanthacris aeriginosa,180
Cyrtotrachelus,154
Dactylispa angulosa,131
Dalceridae,25
Darma sordida,33
Darna,26, 27, 28, 33, 47, 77, 78, 82
Darna bradleyi,25, 32, 47, 78
Darna catenatus,32, 45, 47
Darna diducta,32
Darna furva,32
Darna metaleuca,68
Darna mindanensis,32, 47
Darna nararia,32, 45, 47
Darna pallivitta,33
Darna sordida,45
Darna trima,33, 45, 78, 82
Darna tuaranensis,33
Dasychira inclusa,59
Dasychira mendosa,60, 77
Dedarchis carpophthora,11
Dedarchis euophthalma,11
Dedarchis flavistriata,12
Dedarchis heliotoxa,12
Dedarchis heterogramma,12
Dedarchis minuscula,12
Dedarchis ophiocypha,12
Dedarchis pachygramma,12
Dedarchis psammaula,11

Dedarchis scorpiura,12
Dedarchis spodomicta,12
Delocrania cossyphoides,128, 146
Delphacidae,85, 99, 102
Demotispa aff. pallida,132
Dendubia intemerata,88
Derbidae,84, 85, 105, 108
Derelominae,148
Derelomorphus eburneus,148
Dermestidae,13, 57, 172
Derostenus,138
Devorgilla palmaris,49
Diacrisia strigatula,61
Diacrisia strigatula rubescens,61
Dialeurodes simmondsi,88
Dialomia discreta,147
Dialomia polyphaga,147
Dialox stellatus,204
Diaspididae,89, 95, 97, 98, 108
Diceratothrips bicornis,194
Dichaetothrips brevicollis,194
Diestostemma huallagana,87
Dindymus rubiginosus,27
Diocalandra,157, 161, 162, 165
Diocalandra frumenti,154, 165
Diocalandra impressicolis,154
Diocalandra stigmaticollis,154, 162
Diocalandra taiitensis,154
Dioctes,19, 20
Diostrombus,85
Diptera,7, 8, 9, 13, 17, 24, 25, 27, 29, 30, 51, 52, 53, 55, 56, 60, 61, 66, 82, 140, 154, 163, 180, 184, 185, 186
Discophora celinde,56
Doirania leefmansi,177
Dolicogenidea metesae,8
Dolycotetranychus,199
Drino,184
Dunama angulinea,59
Durrantia,18, 63
Dynamis,152
Dynamis borassi,152, 158, 166
Dynastes hector,119
Dynastinae,112, 119, 120, 121, 123, 124, 125, 126, 211
Dysmicoccus brevipes,91, 104
Dysmicoccus cocotis,91
Echthromorpha agrestoria,8, 48
Echthromorpha tirathabae,48
Elachertidae,53

Elachertus agonoxenae,12, 13
Elachistidae,12
Elaeidiphilos (Epimorius) adustalis,51
Elaeidobius,14, 148, 156, 167, 171
Elaeidobius bilineatus,148
Elaeidobius kamerunicus,14, 148, 167, 171, 194
Elaeidobius plagiatus,148
Elaeidobius singularis,148
Elaeidobius subvittatus,148, 156
Elaeidobius. kamerunicus,148
Elasmidae,9, 16, 19, 48, 54, 138
Elasmus,9, 10, 19, 39
Elasmus albomaculatus,36
Elasmus brevicornis,54
Elasmus fumipennis,48
Elasmus hispidarum,138
Elasmus maculatus,28, 39
Elasmus nephantidis,16, 19, 63
Elasmus philippinensis,54
Elateridae,110, 172
Eldana saccharina,51
Elymnias agondas glaucopis,58
Elymnias agondas glaucopis,74
Elymnias caudata,58
Elymnias fraterna,58
Elymnias hypermestra,58
Elymnias hypermestra undularis,58
Elymnias nesaea,58
Elymnias nigrescens,58
Emersonella palmae,136
Encyrtidae,13, 18, 20, 22, 38, 53, 57, 78, 90, 91, 93, 94, 105, 106, 130, 131, 136, 177
Encyrtus,38
Endomychidae,8
Enicospilus,52
Entedon,138
Entedontidae,53
Entomophthora,180
Entomophthora grylli,180
Eocanthecona,95, 106
Eocanthecona furcellata,43, 95
Eozenillia equatorialis,8
Eoziniella psychidarum,8, 9
Ephestia cautella,49
Episibine intensa,33
Episibine sibinides,33
Epitranus impulsator,48
Epitrimerus elaeis,204

Erechthias flaristriata,10
Erechthias lampadacma,11
Erechthias psammauba,10
Ereunetis ochrocrena,12
Eriborus trochanteratus,16, 19
Erionata thrax,54
Eriophyes guerreronis,200, 202, 203, 205, 206, 207, 208, 210
Eriophyes. guerreronis,202
Eriophyidae,4, 198, 200, 203, 204, 205, 206, 207, 208, 209, 210
Erirhinae,171
Erirhininae,147
Erycia bezzi,53
Ethirothrips brevis,194
Eucarcelia kokiana,49
Euclea cippus,33
Euclea cuprostriga,33
Euclea diversa,33
Euclea norba,33
Euclea plugma,33
Euderastichus,36
Eugnoristus monachus,154
Eugnoristus monachus alluaudi,154
Eulophidae,8, 9, 12, 13, 16, 17, 18, 20, 23, 25, 28, 30, 48, 50, 66, 69, 72, 75, 128, 130, 131, 132, 134, 135, 136, 137, 138, 139, 140, 142, 170, 177
Eumeta,10
Eumolpinae,128, 129
Eupathithrips silvestrii,194
Eupelmidae,9, 19, 23, 28, 36, 37, 53, 138, 184, 186
Eupelmus,13, 37
Eupelmus agonoxenae,13
Eupelmus catoxanthae,8, 9, 23, 36
Eupelmus cushmani,13
Eupelmus javae,23
Euphys bilineata,51
Euplectrini,31
Euplectromorpha,28, 30, 32, 35
Euplectromorpha artonae,23
Euplectromorpha malayensis,34, 36, 38
Euplectromorpha nr. bicarinata,39
Euplectromorpha viridiceps,23
Euplectrus,30, 37, 38
Eupodes,200
Eupodidae,200
Euproctis,63
Euproctis linta,60
Euproctis semisignata,60, 63
Euprosterna eleasa,25, 26, 34, 43, 45

Eupteromalus,13
Eurycantha insularis,185
Euryphlespia cocos,86
Eurytoma,8, 19, 20, 35, 36, 38, 40, 49
Eurytoma albotibialis,17, 19, 20, 23
Eurytoma bracanodis,19
Eurytoma monemae,36
Eurytoma promecothecae,138
Eurytomidae,20, 23, 28, 36, 38, 40, 138
Eurytrachelus,110
Euseius beninensis,199
Eutedon,138
Eutetranychus enodes,199
Exochomus laeviusculus,200
Exorista quadrimaculata,8
Exorista sorbillans,29, 36, 38
Exoristini,29
Exoristobia philippinensis,17, 20
Ferrisia virgata,91
Flatidae,86
Formicidae,7, 18, 49, 82, 106, 196, 197
Fornicia,20, 27, 28, 30, 31, 32, 34, 38, 39, 40
Fornicia annulipes,34, 40
Fornicia ceylonica,27, 30, 32, 36, 38, 39
Fornicia chalcoscelidis,31
Fornicia moronis,40
Fornicia penang,33
Fornicia tagalog,28, 39
Fornicia thoseae,34
Franklinella bruneri,194
Franklinella insularis,194
Franklinella kelliae,194
Fulgoroidea,86, 99
Fulgoromorpha,84, 95, 97
Furcaspis oceanica,91
Furcivena rhodoneuralis,50
Galerucinae,128, 129
Gangara thyrsis,54, 79
Gascardia,91
Gehyra oceanica,178
Gelechiidae,15
Gelechioidea,3, 12, 74
Gelis tenelus,13
Gestronella,131, 139
Gestronella centrolineata,131
Gestronella centrolineata,131
Gestronella lugubris,131
Gestronella lugubris,131

Glyphipterigidae,11, 75
Glyptopanteles,34
Goniozus,18, 19
Goniozus bituberculatus,23
Goniozus fasciatipennis,23
Goniozus inferus,23
Goniozus maculipennis,23
Goniozus nephantidis,16, 19, 20
Goniozus rufobasalis,23
Goriphus varibalteatus,32
Goryphus,16, 31, 38, 39
Goryphus bunoh,9, 38
Goryphus javanicus,23
Goryphus maculiceps,23
Goryphus mesoxanthus,38
Graeffea,78, 186, 187
Graeffea cocophaga,184, 186
Graeffea crouanii,78, 184, 185, 186, 187
Graeffea lifuensis,185
Graeffea seychellensis,185
Gryllidae,180
Grylloidea,4, 180
Gryon,9, 49
Haemishaerota,129, 142
Halictophagous,85, 105
Halticopteroides,129
Haplothrips ceylonicus,194
Hapsifera rugosella,10
Harpagoneura complexa,49
Hectacma indicatrix,12
Hedylepta blackburni,49
Heliodinidae,12
Heliothrips haemorrhoidalis,194
Hemiptera,3, 13, 57, 58, 84, 95
Hemiptera-Cicadomorpha,3, 87
Hemiptera-Fulgoromorpha,3, 84
Hemiptera-Heteroptera,3, 92
Hemiptera-Sternorhyncha,3, 88
Herculia nigrivitta,51
Herminodes insula,59
Hesperiidae,15, 53, 79
Hesperioidea,3, 53
Heteroptera,84, 95, 98, 99, 102, 103, 104, 105, 106, 113
Hexacladia linci,94, 105
Hidari irava,53, 64
Himatidium,128, 141
Himatidium bahianum,128
Himatidium bondari,128

Himatidium neivai,128, 132
Hindsiana arbensis,194
Hindsiana ceylonicus,194
Hindsiana cocois,194
Hirsutella thompsonii,203, 204
Hispinae,28, 128, 130, 132, 133, 135, 136, 139, 140, 143, 144, 145, 211
Hispodiphila brontispae,130, 131
Hispoleptini,130, 136
Hispoleptis,136, 139, 140, 142
Hispoleptis diluta,136
Hispoleptis diluta,136
Hispoleptis elaeidis,136
Hispoleptis ollagnieri,136
Hispoleptis subfasciata,136
Histeridae,10, 113, 154
Homalinotus,157, 163, 167
Homalinotus calvescens,149
Homalinotus coriaceus,149, 157, 163
Homalinotus deplanatus,149
Homalinotus porosus,149
Homoesoma,50
Homoledra sabalella,14
Homophylotis catori,23
Horismenus,35, 128, 170
Hormius,18
Hustachea bondari,147
Hyalospila ptychis,49
Hymenoptera,4, 13, 55, 56, 61, 128, 196
Hypolimnas missipus,58
Hypotermes,188
Hypothenemus tuberculosus,170
Hysteroneura setariae,89
Icerya pattersoni,91
Icerya seychellarum,91
Ichneumonidae,8, 9, 13, 16, 18, 19, 20, 23, 24, 27, 28, 29, 30, 48, 49, 50, 52, 53, 54, 55, 56, 59, 77, 128, 136
Idonauton apicalis,34
Incala setosus,111
Inxwala,86
Iridomyrmex,22
Iridostoma ichthyopa,11
Isdromas,34
Isopedhispa cocotis,131
Isoptera,4, 188
Isyndus heros,8, 27
Kalotermitidae,4, 190
Kamendaka,85

Karnyothrips melaleucus,194
Kermesia,86
Kleothrips subramani,194
Kriechbaumerella ayyari,36
Laelia venosa,60
Laemophloeus ferrugineus,173
Lamoria,51
Lasiodactylus costipennis,171
Lasioderma serricorne,173
Lasioseius,199, 200
Lasioseius phytoseiodes,199
Latoia pallida,26, 34, 45
Latoia viridissima,29, 34, 43, 45, 68, 71, 72, 83
Leefmansia bicolor,177
Lepidoptera,3, 7
Lepidosaphes mcgregori,90
Leptinotarsa decemlineata,128
Leptochirus,113
Leptonatada,61
Leptonatada sjöstedti,60
Leptopharsa gibbicarina,93, 94, 104
Leucocoprinus,196
Leucopholis coneophora,111, 120, 124
Leurostenus elaeidis,155
Levuana iridescens,23, 204
Lienella,38
Limacodidae,7, 25, 26, 27, 28, 29, 30, 43, 45, 46, 47, 63, 64, 65, 66, 67, 68, 69, 70, 71, 73, 77, 80, 81, 95, 211
Lincus,94, 97, 98, 102, 104, 105
Lincus croupius,94
Lincus dentiger,94
Lincus lethifer,94, 104
Lincus lobuliger,94
Lincus malevolus,94, 102, 104, 105
Lincus spurcus,94, 98, 102, 104
Lincus tumidifrons,94, 97
Lincus vandoesburgi,94
Lithargus,171
Litotursus,8
Locris,87
Locusta migratoria manilensis,179
Lophopidae,86, 100, 106
Loxotoma elegans,14, 68
Lucanidae,110
Lupotarsonemus,202
Lydda,85
Lymantriidae,59, 63
Lymeon,18, 34

Lymexilidae,189
Lymexylidae,169
Lyonetiidae,11
Machimia flava,18
Macraspis tetradactyla,111
Macrocentrus,12, 13
Macropygium reticulare,94
Macrotermes,188
Macrotermes bellicosus,188, 191
Macrotermes carbonarius,188
Macrotermes gilvus,188
Macrotermes subhyalinus,188
Mahasena corbetti,7
Manatha albipes,9
Margarodesia desmieri,91
Margarodidae,91, 99
Marpissa,18
Mecodema spinifer,113
Meenoplidae,86
Megacrania phelaus,185
Megalocaria dilatata,89, 100
Megalopyge albicollis,24
Megalopyge lanata,24
Megalopygidae,24
Megaselia,20, 53
Megaselia scalaris,53
Megasoma anubis,119
Melalophacharops papilionis,32
Melanitis leda,58
Melittomma,174, 176, 189, 190
Melittomma insulare,169, 170, 174, 176, 189
Mellitobia sp. nr. hawaïensis,23
Melolonthinae,111, 120
Membracidae,84, 87
Menosca,86, 100
Menosoma elegans,87
Meredolus,148
Meredolus cocois,148
Mesocia dumilla,24
Mesocia pusilla,24
Mesostenus,23
Metamasius,152, 153, 158, 163, 164, 165, 166, 167
Metamasius hemipterus,152, 158, 163, 164, 165, 166
Metaplectrus solitarius,39
Metaplectrus thoseae,34, 40
Metarrhizium,114
Metarrhizium anisophiae,114
Metarrhizium anisopliae,117, 130

Metaxionycha hibrida,129
Metaxionycha testacea,129
Meteoridea hutsoni,17, 19, 20
Meteorus,14, 38
Meteorus trichogrammae,10, 48
Metisa plana,8, 78, 82, 95
Microcerotermes,189, 192
Microcerotermes biroi,188, 192
Microcerotermes bugnioni,189
Microcerotermes los- bañosensis,189
Microcerotermes paraffinis,189
Microcerotermes piliceps,189
Microcerotermes subtilis,189
Microgaster,23
Microporum,171
Microporum congolense,171
Microporum dispar,171
Microstrates bondari,147
Microstrates ypsilon,147
Microthosea celaena,34
Microthosea minima,34, 45, 47
Microthosea porthetes,34
Microtoridea,23
Mimaridae,18
Miridae,93, 99, 103
Molomea,87, 97
Molomea fatalis,87, 97
Molytinae,149
Monolepta,129, 142, 143
Monolepta apicicornis,129, 143
Monolepta limbata,129
Monolepta marica,129
Monolepta tenuicornis,129
Monomorium,18, 134
Monomorium floricola,185
Montronzeriellus melacanthus,58
Mucialla rufivena,48, 51, 81
Mulciber linnaei,110, 171
Mycetophagidae,171
Mycteromyiella,186
Mycteromyiella laetifica,185
Mycteromyiella phasmatophaga,185
Myllocerus curvicornis,154, 161, 165
Mymaridae,177
Myndus,85, 97, 98, 100, 101, 103, 108
Myndus adiopodoumeensis,86
Myndus chazeaui,86
Myndus crudus,85, 98, 100, 101

Myndus macfarlanei,86
Myndus mavors,86
Myndus rumina,86
Myndus taffini,86, 101, 103
Myrmecolacidae,177
Mysteroneura setariae,89
Mystrops,148, 171
Mystrops costaricensis,171
Mystrops costaricensis costaricensis,171
Mystrops costaricensis orientalis,171
Mystrops costaricensis pacificus,171
Nacerimina gutierrezi,204
Nacerimina jamaica,204
Narosa rosipuncta,34
Nasutitermes,189
Nasutitermes ceylonicus,189
Nasutitermes costalis,189
Nasutitermes ephratae,189
Nasutitermes havilandi,189
Nasutitermes laticeps,189
Nasutitermes lugonicus,189
Nasutitermes matangensiformis,189
Nasutitermes negritus,189
Nasutitermes novarumhebridarum,189
Nasutitermes princeps,189
Nasutitermes ripertii,189
Natada fusca,34
Natada michorta,35
Natada pucara,35, 45
Natada subpectinata,35, 47
Nealsomyia rufella,8, 29, 31
Neasteropaeus,17, 20
Necrobia rufipes,172, 174
Neoaplectana carpocapsae,114
Neochryopus savagei,111, 113
Neolecanium silverai,92
Neoplectrus,28, 30
Neoplectrus arcolatus,32
Neoplectrus bicarinatus,23
Neoplectrus clavatus,32
Neoplectrus maculatus,23, 32
Neoplectrus nr. clavatus,32
Neoseiulus,202
Neoseiulus baraki,202
Neoseiulus mumai,202
Neoseiulus paspalivorus,202
Neotelenomus javae,53
Nephantis serinopa,15, 23, 67, 68, 71, 75, 76, 77, 79, 204

Nesodryas antiope,85
Nesolynx,23, 28, 32, 36, 37, 136
Nesolynx zygaenarum,23
Nesothrips brevicollis,194, 195
Neuroptera,17
Nipaecoccus nipae,91
Nitidulidae,148, 171, 172
Noctuidae,58, 63, 83
Noctuoidea,3, 58
Nodocnemus uniformis,147
Nolinae,59
Norape,25
Norape aff. camela,24
Norape argyrrhorea,25
Notodontidae,60, 67, 81
Notogalumna nortoni,204
Notolomus,148
Notolomus elaeisae,148
Notolomus mariae-helenae,148
Notostrix attenuata,204
Numicia,86
Nylanderia fulva,13, 196, 197
Nymphalidae,57, 74, 76
Nzinga,87, 109
Ochlerus,94
Octodonta afinis,131
Odites,15
Odontomachus,22
Odontomachus simillimus,185
Odontotermes,188, 193
Odontotermes horni,188
Odontotermes obesus,188
Odontotermes obsauriceps,188
Odontotermes redemanni,188
Oecophoridae,15, 63, 65, 204
Oecophylla,93, 196
Oecophylla longinoda,92, 104, 106, 108, 134
Oecophylla smaragdina,18, 49
Oencyrtus,78, 130
Oiketicus,73, 78
Oiketicus abbotti,10
Oiketicus kirbyi,9, 73, 78
Oligonychus,198, 199, 205, 206, 210
Oligonychus bagdasariani,199
Oligonychus indicus,198
Oligonychus iseilemae,198, 200, 210
Oligonychus modestus,199
Oligonychus plegas,199

Oligonychus pratensis,199
Oligonychus velascoi,198, 205
Oligosita,134, 136, 146
Oligosita longiclavata,135
Oligosita minuta,134
Oligosita robusta,134
Oligosita utilis,138
Oligota pallidicornis,200
Olona,26
Olona albistrigella,35
Olona gateri,35
Omolicna,85
Omotemnus miniatocrinitus,154
Oncometopia,87
Oncometopia venata,87
Ooencyrtus,13, 53, 57, 106, 131
Ooencyrtus af. calpodicus,22
Ooencyrtus podontiae,131
Ooencyrtus utetheisae,93, 106
Ootetrastichus dubius,177
Ophicrania leveri,185
Ophioninae,52
Ophryocystis oryctesi,115, 125
Opisina arenosella,15, 16, 17, 19, 20, 63, 65, 69, 71, 74, 75, 76, 77, 79, 81, 82, 83, 204
Opogona aurisquamosa,11
Opogona dimidiatella,11
Opogona leucodelta,11
Opogona pileigera,11
Opogona purpuriella,10
Opogona regressa,11
Opogona taochroa,11
Opogona trigonomis,11
Opsiphanes cassiae,57
Opsiphanes cassina,56, 64
Opsiphanes invirae,57
Opsiphanes quiteria,57
Opsiphanes tamirindi,57
Orgyia turbata,60, 65
Oribatidae,4, 204
Ormenaria rufifascia,86
Orthoptera,4, 177
Oryctes,95, 107, 112, 113, 114, 115, 116, 117, 118, 120, 121, 122, 123, 124, 125, 126, 127, 172
Oryctes augias,112
Oryctes boas,107, 112, 113, 115
Oryctes centaurus,112
Oryctes chevrolati,112

Oryctes comoriensis,112
Oryctes erebus,112
Oryctes gigas,112
Oryctes insularis,112
Oryctes monoceros,112, 113, 115, 116, 122, 123, 125, 126
Oryctes preussi,119
Oryctes pyrrhus,112, 113, 118
Oryctes rhinoceos,112
Oryctes rhinoceros,107, 112, 113, 114, 115, 116, 120, 121, 122, 123, 124, 125, 126, 127
Oryctes simiar,112, 113
Oryctoderus latitarsis,119
Oryzaephilus mercator,172
Oryzaephilus surinamensis,172
Oxya velox,179
Oxya vicina,179
Oxycatantops spissus,180
Oxycephala,131
Oxycephala cornigera,131
Oxycephala wallacei,131
Oxycephala. papuana),131
Oxydema fusiforme,155
Pachnoda marginella,111
Pachylister chineensis,113
Pachymerus cardo,173, 176
Pachymerus lacerdae,173, 174
Padraona chrysozona,54
Paecilomyces,47, 55
Paecilomyces farinosus,18
Paecilomyces javanicus,47
Painella simmondsi,86
Palaecorynus cerinipes,147
Palexorista,29, 38
Palexorista coccyx,29
Palexorista inconspicuoides,53
Palexorista nr. mayneana,29, 34
Palexorista nr. ophirica,29
Palexorista painei,48, 49
Palexorista solennis,8, 29, 32
Palmaspis,92
Palmaspis palmae,92
Palmaspis ucayali,92
Palmicultor palmarum,91
Palmistichus elaeisis,18, 28, 39
Palpexorista coccyx,39
Pamendanga,85
Papilionoidea,3, 55
Papuana armicollis,119

Papuana hübneri,119
Papuana loevipennis,119
Papuana semistriata,119
Papuana woodlarkiana,119
Parabillea rhynchophorae,150
Paradasynus rostratus,92, 107
Paraleyrodes,88
Paranastatus,186
Paranastatus nigriscutellatus,184
Paranastatus verticalis,184
Paraphylax rufipes,9
Paraphylax sorianoi,40
Paraphylax varius,8, 32
Parasa,25, 27, 35, 47, 64, 66, 69, 73, 76
Parasa balitkae,35
Parasa chlorostigma,35
Parasa corbetti,35
Parasa darma,35
Parasa darmoides,35
Parasa lepida,28, 29, 36, 43, 45, 47, 69, 76
Parasa philepida,36, 47
Parasa sundalepida,36
Parasierola,9
Paratheresia brasilensis,153
Paratheresia menezesi,150
Parena laticincta,17, 68
Parena nigrolineata,17, 76, 77
Parisoschoenus,147
Parisoschoenus obesulus,147
Parisoschoenus suturalis,147
Parlagena bennetti,91
Passalidae,110
Patanga luteicornis,179
Patanga succincta,179
Pectinarosa alastor,36
Pediobius,8, 9, 16, 19, 36, 50, 69, 71, 131
Pediobius anomalus,8, 138
Pediobius balyanae,135
Pediobius brunchicida,16, 19
Pediobius coffeicola,134
Pediobius detrimentosus,19, 23
Pediobius elasmi,23, 35, 36, 37
Pediobius imbreus,8, 16, 17, 19, 20, 69, 131
Pediobius lividiscutum,8, 9
Pediobius parvulus,23, 136, 138
Pediobius ptychomyiae,23, 35, 36
Pediobius setigerus,134
Pediobius vigintiquinque,134

Peleopoda,18
Peleopoda arcanella,18
Pendlury,56
Pentatomidae,8, 13, 27, 43, 57, 58, 93, 94, 95, 97, 98, 102, 104, 105, 106, 136
Penthocrates,25, 26, 27, 28, 36, 47, 65
Penthocrates albicapitata,36
Penthocrates bigenita,37
Penthocrates rufa,37
Penthocrates rufofascia,37, 47
Penthocrates styx,37, 47
Penthocrates zelaznyi,37
Peregrinus maidis,85
Pericapritermes,188
Perichares corydon,54
Perilampidae,48
Perilampus,48, 49
Perilampus microgastris,19, 48
Perilitus persimilis,134
Periphoba hircia,52, 66
Perisierola nephantis,63
Pestalotiopsis,10, 25
Phalacridae,8
Phasmida,4, 184
Phasmidae,78, 184
Pheidole,22, 93, 104
Pheidole megacephala,93, 104, 134, 185
Phillocoptes mariaui,204
Phlaedromius nigrolineatus,17
Phobetron,28
Phobetron hipparchia,37
Phoridae,20, 51, 53
Phyleurus castaneus,119
Phyleurus didymus,119
Phytometra acuta,59
Phytoseiidae,199, 202
Picornaviridae,61
Pimelephila ghesquieri,50
Pimelopus tenuestriatus,119
Pimetephila ghesquieri,69
Pimplinae,52, 136
Pinnaspis strachani,91
Pistosia testacea,131
Plaesius javanus,154
Planococcus lilacinus,47, 91
Platyauchenia,132, 144
Platyauchenia deyrollei,132
Platyauchenia latreillei,132, 144

Platygenia barbata,111, 121
Platymerus,107, 113
Platymerus biguttata,95, 114
Platymerus laevicollis,113
Platymerus rhadamanthus,95, 107, 113
Platynopus,95
Platyplectrus,28, 31, 32
Platyplectrus orthocraspedae,32, 33
Platypodidae,170, 175
Platypus rugulosus,170, 175
Pleistophora tanzaniae,115, 125
Pleseobyrsa bicincta,93, 94, 104
Plesispa,131
Plesispa papuana,131
Plesispa reichei,131
Plesispa ruficollis,131
Pleurotropis,19
Pleurotroppopsis lankensis,138
Pleurotroppopsis podagrica,134, 135
Plötzia amygdalis,55
Pochazia fasciata,86
Podalia bolivari,24
Podissus,57
Podissus agenor,27
Poophilus,87
Praestochrysis shanghaiensis,28
Prionacis ligaoi,36, 37
Pristhesancus femoralis,58
Pritchardina fijiensis,198
Proarna hilaris,88
Prodenia litura,59
Promecotheca,135, 137, 138, 140, 141, 142
Promecotheca callosa,135
Promecotheca coeruleipennis,135, 136, 138, 204
Promecotheca cumingi,135, 136, 138, 140, 141
Promecotheca lindingeri,135
Promecotheca nuciferae,135
Promecotheca opacicollis,135, 138, 142
Promecotheca papuana,135, 138
Promecotheca reichei,135, 138
Promecotheca soror,135
Promecotheca varipes,135
Promecothecini,130, 135
Prosapegus atrellus,177
Prosoestus minor,148
Prosoestus sculptilis,148
Prosoestus sculptilis,148
Protocerius colossus,154

Proutista,85, 100, 105
Proutista moesta,85, 100, 105
Psenoflata brevis,86
Pseudaspidoproctus vaysieriellus,91
Pseudobrachymeria,13, 34
Pseudocentrinus lucidulus,147
Pseudococcidae,91, 108
Pseudococcus,91, 104
Pseudococcus adonidum,91
Pseudococcus citriculus,91
Pseudococcus cryptus,91
Pseudococcus longispinnus,91
Pseudocossus uliginosus,21
Pseudodirphia gragatus,52
Pseudolacapsis,91
Pseudolacapsis cockerelli,91
Pseudometopia,87
Pseudometopia amblardii,87
Pseudonirmides sola,37
Pseudoscymnus,90
Pseudoscymnus anomalus,90
Pseudotheraptus,92, 95, 97, 102, 104, 106, 107
Pseudotheraptus devastans,92, 95
Pseudotheraptus wayi,92, 93, 97, 104, 106, 107
Pseudotrochatus concolor,111
Pseudotrochatus nigroviridis,111
Psilopholis vestita,111, 126
Psychidae,7, 78, 82
Psyttala dudgeonia,95, 114
Pterolophia apiceplagiata,171, 174
Pteroma,9
Pteromalidae,13, 14, 129
Pteroteinon laufella,54, 55, 70
Pullus xerampelinus,90, 106
Pyemotes ventricosus,18, 19, 20, 136, 204
Pyemotidae,4, 150, 204
Pygmaeomorpha,38
Pygmaeomorpha priori,37
Pygmaeomorpha ramosa,37
Pyralidae,48, 68, 69, 81, 83, 127, 173
Pyraloidea,3, 48, 83
Pyrgomophidae,179
Pyrgomorphidae,177, 178, 179, 181, 182
Pyroderces,14
Pyroderces amphisarsis,14
Pyrrhochalcia iphis,55
Pyrrhocoridae,27
Quasithosea sythoffi,38

Raoiella indica,199, 200, 209
Rarosiella cocosae,199, 209
Recilia mica,87, 98
Reduviidae,8, 9, 13, 27, 58, 95, 113, 114
Retracrus elaeis,203, 206
Rhabditidae,11, 150, 158
Rhabditis adenobia,117, 125
Rhabdoscelus,156
Rhabdoscelus asperipennis,154
Rhabdoscelus lineatocollis,154
Rhabdoscelus obscurus,153, 156
Rhagionidae,154
Rhinostomus,157, 167
Rhinostomus afzelii,153
Rhinostomus barbirostris,153, 157
Rhinostomus niger,153
Rhinostomus scrutator,153
Rhinotermitidae,4, 189, 192
Rhizobius lophanthae,90
Rhizoecus,91, 104
Rhizoecus nr. americanus,91
Rhogas,34, 36
Rhopalocampta bixae,55
Rhopalosiphum rufiabdominalis,89
Rhynchophorinae,149, 166, 167
Rhynchophorus,120, 149, 150, 151, 152, 156, 157, 158, 159, 160, 161, 162, 163, 164, 165, 166, 167, 168
Rhynchophorus bilineatus,149, 151, 164
Rhynchophorus distinctus,149
Rhynchophorus ferrugineus,150
Rhynchophorus ferrugineus,149
Rhynchophorus ferrugineus,149, 150, 151, 152, 156, 157, 158, 159, 160, 161, 163, 165, 167
Rhynchophorus lobatus,149
Rhynchophorus palmarum,151, 152
Rhynchophorus palmarum,150
Rhynchophorus palmarum,149, 150, 151, 152, 156, 158, 159, 160, 161, 162, 163, 164, 166
Rhynchophorus phoenicis,150, 151, 153, 159, 163, 164, 166
Rhynchophorus quadrangalus,150
Rhynchophorus ritcheri,149
Rhynchophorus vulneratus,149, 150, 151, 160, 162, 163, 165
Rhynchotarsomemus,202
Rhysipolis,13
Rhyzobius satelles,90
Ricania speculum,86
Ricaniidae,86
Rogas,31, 32, 33, 34, 38, 39

Ropalidia nigra,49
Rutelinae,110
S. decoratus,178
Sacila variipes,8
Sagalassa valida,11, 63, 68, 75
Saissetia lutea,91
Saliana,54
Saliana severus,54
Salticidae,18, 79, 184
Sangicoccus sp. nr. truncatispinus,92
Sarcolosomia crinitus,36
Sarcophaga antilope,36, 53
Sarcophaga exuberans,180
Sarcophaga fuscicauda,53
Sarcophaga orientaloides,53
Sarcophagidae,30, 70, 180, 184
Sarcorohdendorfia antilope,31
Sarcosolomia crinita,35
Saturniidae,52, 66, 68
Scapanes,112, 117, 120, 121, 123, 124, 125, 126
Scapanes australis,116, 120, 121, 123, 124, 125, 126
Scapanes australis australis,117
Scapanes australis brevicornis,117
Scapanes australis grossepunctatus,117, 120
Scapanes australis salomonensis,117
Scaptocoris,94, 97, 108
Scaptocoris divergens,94, 108
Scaptocoris terginus,94
Scarabaeidae,3, 95, 110, 112, 116, 117, 211
Scarites madagascariensis,113
Scelionidae,9, 48, 49, 53, 55, 56, 57, 59, 61, 177
Scenocharops punctata,32, 40
Schedorhinotermes,190
Schedorhinotermes longirostris,190
Schedorhinotermes malaccensis,190
Schedorhinotermes marjoriae,190
Schistocerca flavofasciata,179
Schistocerca paranensis,179
Schizonycha africana,111
Schizotetranychus hindustanichus,198
Schzaphis rotundiventris,89
Scolia,113, 126, 156
Scolia erratica,150, 156
Scolia oryctophaga,113
Scolia procer,113
Scolia ruficornis,113, 126
Scoliidae,113, 150
Scolocenus spiniferous,204

Scolytidae,169, 175
Scotothrips claripennis,194, 195
Segestes,178, 183
Segestes beieri,178
Segestes brevipennis,178
Segestes celebensis,178
Segestes cornelli,178
Segestes decoratus,177, 178, 183
Segestes frater,178
Segestes fuscus,178
Segestes punctipes,178
Segestes stibicki,178
Segestes unicolor,178
Segestes vittacipes,178
Segestidea,178, 182
Segestidea acuminata,178
Segestidea gracilis,178
Segestidea gracilis gracilis,178
Segestidea gracilis simulatrix,178
Segestidea leefmansi,178
Segestidea marmorata,178
Segestidea marmorata marmorata,178
Segestidea marmorata occidentalis,178
Segestidea novaeguineae,178
Segestidea punctipennis,178
Segestidea rufipalpius,178
Segestidea soror,178
Segestidea uniformis,177, 178, 182
Seniorwhitea krameri,31
Seniorwhitea orientalis,31, 36, 38
Seratia marcescens,18
Sericopimpla flavobalteata,9
Sesamia,63, 71
Sesamia calamistis,58
Sesamia inferens,58, 71
Sesamia poephaga,58
Sesioidea,3, 21
Setora,25, 26, 27, 74, 79, 80, 81
Setora cupreistriga,38, 47
Setora kelapae,38
Setora longifurca,38, 45
Setora nitens,25, 26, 27, 29, 38, 46, 47, 74, 79, 80, 81
Setora tagalog,38, 46
Setothosea asigna,25, 26, 27, 29, 39, 43, 46, 47, 80
Sexava,177, 181, 182, 183
Sexava coriacea,177, 178
Sexava karnyi,178
Sexava nubila,178

Sibine,25, 26, 27, 28, 33, 39, 68, 69, 74, 95, 100
Sibine apicalis,39, 43, 46
Sibine fusca,28, 29, 33, 39, 43, 46, 68, 69, 74, 95, 100
Sibine megasomoides,39
Sibine modesta,39
Sibine nesea,39, 43, 46
Sibine pallescens,39
Sibine trimaculata,39
Siculobata malabarica,204
Smicrips af.exilis,171
Sochinsogonia longa,87
Sogatella,102
Sogatella kolophon,85
Solenopsis geminata,185
Sophonia greeni,87, 105
Sophrops eurystoma,111
Spaethiella costipennis,129
Spaethiella tristis,128, 136
Sparganobasis subcruciatus,154
Spathius apicalis,154
Spilochalcis,34, 39, 56, 170
Spilochalcis nigrifrons,57
Spilochalcis nunani,34
Spinaria,27, 39
Spinaria armata,34, 40
Spinaria spinator,27, 38
Spodoptera,58, 83
Spodoptera exempta,59, 83
Spodoptera litura,59
Spodoptera mauritia,59
Sporothrix insectorum,94, 104
Staphylinidae,10, 113
Stegonotus modestus,178
Steinernema carpocapsae,11
Steinernema feltiae,114
Steinernema glaseri,117
Steinernematidae,11
Steneotarsonemus furcatus,200, 202
Stenispa,132
Stenoma cecropia,13, 68
Stenomatinae,7, 13
Stenomesius,34
Stenotrupis filum,155
Stephanitis,101, 103, 105
Stephanitis typica,93, 105
Stephanoderes,170
Stephanoderes perhispidus,170
Stephanoderes punctatus,170

Sternorhyncha,84
Stethoconus praefectus,93, 103
Stethorus,200, 206
Stethorus expectatus,200, 206
Stethorus exsultabilis,200
Stethorus keralicus,200
Stethorus parcepunctatus,200
Stethorus siphonulus,200
Stethorus tetranychii,200
Stethorus vinsoni,200
Stethynium,18
Sthathmopoda adulatrix,12
Sthathmopoda nucivora,12
Stichotrema dallatorreanum,177
Stictopisthus,13, 36, 40
Stigmatidae,200
Stilbella,134
Stodiplosis gagnei,148
Stomatocerus,54
Stomatocerus sulcatiscutellum,16
Stomatomyia bezziana,17, 20
Strategus,112, 113, 117, 118, 123, 125, 126
Strategus aloeus,118, 123, 126
Strategus anachoreta,118
Strategus barbigerus,118
Strategus quadrifoveatus,118
Strategus simson,118
Strategus validus,118, 125
Struthocelis semiotarsa,18
Sturmiini,29
Suastus gremius,53, 79
Sufetula,50, 51, 66, 68
Sufetula diminutalis,50, 68
Sufetula minimalis,51
Sufetula nigrescens,50
Sufetula sunidesalis,51, 66
Sukunahikona popei,90, 107
Susica malayana,28, 39
Syagrus puncticollis,129
Sycanus,8, 9, 27, 95
Sycanus dichotomus,27
Sycanus leucomesus,27
Sycanus macracanthus,8, 27
Symphiothrips punctatus,194
Sympiesis,8
Sympiesis aburiana,134, 135
Sympiesis javanica,138
Synale hylaspes,54

Syntermes,189, 193
Syntomosphyrum obscuripes,20, 23
Syntomosphyrum.obscuripes,23
Syrphidae,18
Systropus,29
Systropus nitidus,29, 39
Systropus pelopeus,34
Systropus polistoides,29
Systropus roekpei,33
Systropus roepke,34
Systropus roepkei,30, 31
Tachina munda,36
Tachinidae,8, 9, 13, 14, 17, 20, 23, 24, 25, 28, 29, 30, 48, 49, 50, 52, 53, 54, 55, 56, 60, 61, 64, 66, 70, 71, 82, 154, 163, 184, 185, 186
Tachinobia repanda,23, 25
Taeneris,58
Taeneris myops,58
Tagasodes cubanus,85
Tambinia,86
Tambinia verticalis,86
Taphrocerus cocois,170
Tapinoma melanocephalum,91, 185
Tarsonemidae,202
Tarsonemus,202
Technomyrmex albipes,18
Tegonotus gutierrezi,204
Telenomus,9, 49, 56, 57
Telenomus thoas,55
Telenomus tirathabae,48, 49
Telenomus urios,53
Telicota,54, 76
Telicota bambusae,54
Telicota palmarum,54, 76
Temnochoita basipennis,153
Temnorrhynchus coronatus,119
Temnoschoita delumbata,153
Temnoschoita erudita,153
Temnoschoita maynei,153
Temnoschoita quadripustulata,153
Temnoschoita subulirostris,153
Tenebrionidae,110, 173
Tenuipalpidae,4, 199, 209
Terathorabditis palmarum,150
Termes,189
Termes carbonavius,189
Termes classicus,189
Termitidae,4, 188, 192, 193
Tetramorium aculeatum,134

Tetranychidae,4, 198, 199, 200, 206, 207, 209, 210
Tetranychus,198, 206, 207, 209
Tetranychus fijiensis,198
Tetranychus hindustanicus,198
Tetranychus ludeni,198, 209
Tetranychus mexicanus,198
Tetranychus neocaledonychus,198
Tetranychus piercei,198, 199
Tetranychus tumidus,198, 207
Tetrapolipus rhynchophori,150
Tetrastichidae,23
Tetrastichus,9, 13, 25, 34, 40, 54, 75, 132, 136, 140, 177
Tetrastichus brontispae,130, 140
Tetrastichus israeli,17, 20, 75
Tetrastichus plesispae,131
Tetrastichus sp. nr. dubius,177
Tettigoniidae,177
Tettigonioidea,4, 177, 182, 183
Thelairodrino gracilis,20
Theronia,34
Theronia orientalis,38, 39, 40, 55
Thosea,25, 27, 29, 46, 47, 63, 70, 71, 73, 81
Thosea andamanica,40
Thosea austrocaerulea,40
Thosea bipartita,40
Thosea boreocaerulea,40, 46
Thosea chrysoparala,40
Thosea lutea,40, 47
Thosea mediostrigata,40
Thosea monoloncha,40, 46, 73
Thosea phillipina,40
Thosea siamica,41
Thosea unifascia,41, 71
Thosea vetusta,41
Thratala,20
Thrips hawaiiensis,194
Thysanoptera,4, 194
Tiarothrips subramanii,194
Tilachlidiopsis,134
Tillancoccus,92
Tineidae,10, 59
Tineoidea,3, 7
Tingidae,93, 100, 103, 104
Tiquadra,59
Tiquadra circumdata,10
Tirathaba,49, 69, 76, 83, 127
Tirathaba complexa,48
Tirathaba fructivora,48

Tirathaba mundella,48, 83
Tirathaba rufivena,48, 49, 127
Tonesia melas,147
Tongamyia cinerella,13
Trachycentra calamias,10
Trachycentra elaetropha,10
Trachysphyrus oxymorus,36
Trathala flavoorbitalis,13, 49
Tribolium castaneum,173
Trichiinae,111
Trichogornphus semmelinki,119
Trichogornphus vicinus,119
Trichogramma,14, 49
Trichogramma brasilensis,17
Trichogramma minutum,17
Trichogramma pretiosum,13
Trichogrammatidae,13, 28, 30, 48, 128, 134, 136, 138, 146, 177
Trichogrammatoidea,41
Trichogrammatoidea nana,48, 130
Trichogrammatoidea thoseae,28, 32, 38, 39, 40
Trichogyia semifascia,41
Trichohalticella tirathabae,49
Trichomalopsis,13
Trichospilus,50
Trichospilus pupivora,16, 17, 20, 48, 49
Tricogrammatidae,138
Trighogrammatidae,130
Trigonata hyalinata var. branneri,196
Trilophidia annulata,179
Triphleps,17
Triraphis,38
Triteleia atrella,177
Trogocrada deleter,41
Trogoderma granarium,172
Tropidacris,179
Tropidacris collaris,179
Tropidacris cristata grandis,179
Tropidacris cristata latreilli,179
Tropiduchidae,86
Trypanosomatidae,94
Trypoxylon,13
Turnaca,61, 67, 81
Turnaca acuta,61, 81
Turnaca rufisquamata,60, 67
Tydeidae,202
Typhlocibinae,87, 109
Typhlodromus caudatus,200
Uracrobates indicus,204

Valanga,179, 181
Valanga nigricornis,179, 181
Valanga nigricornis sumatrensis,179
Valanga transiens,179
Valanga zehntneri,179
Verticillium lecanii,47
Vespidae,49
Vespoidea,113
Vinsonia stellifera,91
Virgilia,86
Winthemia,20
Xanthopimpla,8, 9, 20, 38, 54, 77
Xanthopimpla gampsura,53
Xanthopimpla nana,20
Xanthopimpla pedator,9
Xanthopimpla punctata,20
Xanthopimpla regina,53
Xanthozona melanopyga,56, 82
Xyleborus,170, 175
Xyleborus affinis,170, 175
Xyleborus confusus,170
Xyleborus ferrugineus,170, 175
Xyleborus perforans,170
Xyleborus similis,170, 175
Xyleborus torquatus,170
Xylorictidae,15, 75, 76, 81
Xylotrupes,119, 121
Xylotrupes gideon,119
Xylotrupes lorquini,119
Yponomeutoidea,3, 11
Zonocerus variegatus,179, 180, 181, 182, 183
Zophiuma,86, 106
Zophiuma lobulata,86, 106
Zophopetes cerymica,55
Zorabana,85
Zoraïda,85
Zygaenidae,17, 22, 204
Zygaenoidea,3, 22

Mahasena corbetti: larva in its case.
(R. Desmier de Chenon)

Agonoxena argaula: damages.
(J.P. Morin)

Opisina arenosella: larva. (P.N. Fernando)

Sagalassa valida: larva *in situ.*
(Ph. Genty)

Loxotoma elegans: case. (D. Mariau)

Peleopoda arcanella: adult. (D. Mariau)

Castnia dedalus: adult. (D. Mariau)

Castnia dedalus: larva. (D. Mariau)

Castnia dedalus: damage on a bunch. (D. Mariau)

Artona catoxantha: larva. (R. Desmier de Chenon)

Homophylotis catori: larva.
(D. Mariau)

Norape argyrrhorea: larva.
(Anon.)

Mesocia pusilla: larva and damage.
(D. Mariau)

Phobetron sp.: larva. (Ph. Genty)

Darna bradleyi: larva and damage.
(R. Desmier de Chenon)

Palms affected by *Darna bradleyi*.
(R. Desmier de Chenon)

Parasa lepida: larva.
(R. Desmier de Chenon)

Setora nitens: larvae. (D. Mariau)

Setora nitens: larva affected by virus.
(R. Desmier de Chenon)

Oilpalms affected by *Setora nitens*.
(D. Mariau)

Parasa philepida: larva. (M.J.W. Cock)

Episibine intensa: larvae healthy and attacked by *Apanteles* sp. (Ph. Genty)

Sibine sp.: colony of young larvae. (D. Mariau)

Sibine fusca: larvae and damages. (Ph. Genty)

Natada fusca: larva. (Ph. Genty)

Euclea cippus: larva. (Ph. Genty)

Euprosterna elaeasa: healthy (right) and affected larva by a poliedrosis. (Ph. Genty)

Larva of *Spinaria spinator* inside a larva of *Setora nitens.*
(R. Desmier de Chenon)

Casinaria sp.: adult. (Ph. Genty)

Chaetexorista javana going out from a pupa of *Limacodidae*.
(R. Desmier de Chenon)

Tirathaba rufivena: larva and damages.
(R. Desmier de Chenon)

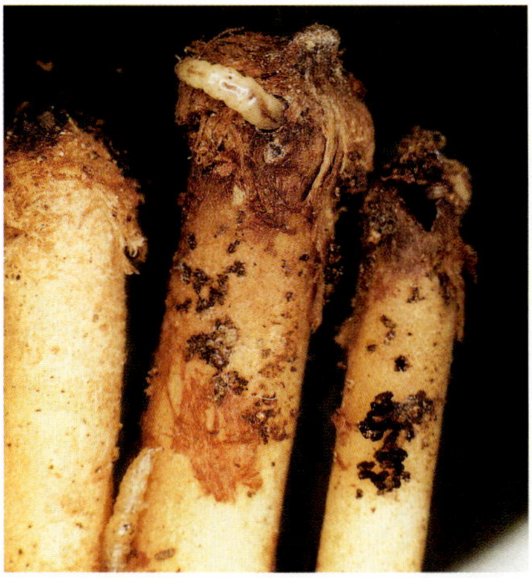

Sufetula diminutalis: larva and damage.
(Ph. Genty)

Sufetula sunidesalis: root affected.
(D. Mariau)

Pimelephila ghesquieri: larva and damage. (D. Mariau)

Hidari irava: larva.
(R. Desmier de Chenon)

Brassolis sophorae: eggs parasitized by *Telenomus* sp. (D. Mariau)

Brassolis sophorae: larvae.
(D. Mariau)

Brassolis sophorae: damages. (Ph. Genty)

Opsiphanes cassina: larva.
(D. Mariau)

Opsiphanes cassina: pupa.
(Ph. Genty)

Aleurodicus coccois: adults.
(M. J. W. Cock)

Aleurodicus sp.: larvae. (Ph. Genty)

Aspidiotus destructor with adults and pupae of ladybird. (J.P. Morin)

Leptopharsa gibbicarina and *Pestalotiopsis* attacks. (R. Desmier de Chenon)

Pseudotheraptus devastans: adult. (D. Mariau)

Pleseobyrsa bicincta: adult. (CIRAD)

Lincus croupius: adult. (J.P. Morin)

Oryctes rhinoceros: female.
(R. Desmier de Chenon)

Oryctes rhinoceros: larva.
(R. Desmier de Chenon)

Coconut palm frequently attacked by *Oryctes monoceros*. (D. Mariau)

Oryctes rhinoceros gallery on a young oil palm. (D. Mariau)

Scapanes australis: male.
(R. Desmier de Chenon)

Strategus aloeus: male. (Ph. Genty)

Demotispa sp.*:* adult. (Ph. Genty)

Strategus aloeus attack on coconut. (D. Mariau)

Coconut attacked by *Augosoma centaurus.* (D. Mariau)

Coelaenomenodera lameensis: adults. (D. Mariau)

Coelaenomenodera lameensis: larva and pupa. (D. Mariau)

Oil palms affected by *Coelaenomenodera lameensis*. (D. Mariau)

Hispoleptis sp.: larval gallery. (D. Mariau)

Brontispa longissima: adult. (J. P. Morin)

Young coconut palm affected by *Brontispa longissima*. (D. Mariau)

Coconut palm highly affected by *Gestronella centrolineata.*
(D. Mariau)

Rhynchophorus palmarum: adult.
(Ph. Genty)

Coraliomella brunnea: adult.
(J. P. Morin)

Young oil palm attacked by *Temnoschoita quadripustulata.*
(D. Mariau)

Rhabdoscelus obscurus: larva *in situ.*
(J.F. Julia)

Amerris ynca: adult. (J.I.L. Moura)

Sexava coriacea: female. (R. Desmier de Chenon)

Coconut defoliated by *Segestes decoratus.* (L. Beaudoin-Ollivier)

Valanga nigricronis: adult. (R. Desmier de Chenon)

Graeffea crouanii: adult. (G. Rapp)

Atta cephalotes: damage.
(D. Mariau)

Coptotermes curvignathus:
young oil palm affected.
(R. Desmier de Chenon)

Coptotermes curvignathus: nest.
(J. Renoux)

Macrotermes gilvus: nest.
(R. Desmier de Chenon)

Macrotermes bellicosus:
young coconut palm attacked.
(D. Mariau)

Eriophyes guerreronis: young coconut bunch highly attacked. (D. Mariau)

Retracrus elaeis: leaflets attacked. (Ph. Genty)

Tetranychidae: oil palm attacked in nursery. (D. Mariau)